RED© LIST

THE IUCN RED LIST
OF THREATENED SPECIES™

SPECIES ON THE EDGE OF
SURVIVAL

SPECIES ON THE EDGE OF SURVIVAL

Published jointly by the IUCN, International Union for Conservation of Nature
28, Rue Mauverney, 1196 Gland, Switzerland,
and HarperCollins Publishers Ltd.
Westerhill Road, Bishopbriggs, Glasgow G64 2QT.

First edition 2011

ISBN 978-0-00-741914-2

Printed and bound in China

The ideas and opinions expressed in this publication are those of the authors and are not
necessarily those of IUCN and do not commit the Organization.
The designations employed and the presentation of material throughout this publication
do not imply the expression of any opinion whatsoever on the part of IUCN concerning
the legal status of any country, territory, city or area or of its authorities or concerning
the delimitation of its frontiers or boundaries.

For more information on IUCN, please contact:
IUCN, International Union for Conservation of Nature
28, Rue Mauverney
1196 Gland, Switzerland
e-mail: **species@iucn.org**
www.iucn.org

Cover images
Front: Hawksbill Turtle (Rich Carey), Asian Elephant (A v.d. Wolde),
Franklin's Bumble Bee (Pete Schroeder), Luristan Newt (Twan Leenders),
Back: Inthanon Nat. Park (Image Focus), Antsingy Leaf Chameleon (Richard K. B. Jenkins/Madagasikara)

While every effort has been made to trace the owners of copyright material reproduced herein
and secure permissions, the publishers would like to apologise for any omissions and will be
pleased to incorporate missing acknowledgements in any future edition of this book.

All mapping in this book is generated from Collins Bartholomew digital databases.
Collins Bartholomew, the UK's leading independent geographical information supplier,
can provide a digital, custom, and premium mapping service to a variety of markets.
For further information:
Tel: +44 (0) 141 306 3752
e-mail: **collinsbartholomew@harpercollins.co.uk**
or visit our website at: **www.collinsbartholomew.com**

In today's world, the diversity of animals, fungi and plants that exist are the product of 3.5 billion years of evolution. This evolutionary process has been affected by the impacts of people over the last 1-2 million years and in particular over the last 500 years. Current estimates of the number of species range from 5 to 30 million, with a best estimate of 8 to 14 million; of these, only around 1.8 million have been named or 'described'.

While scientists debate how many species exist, there are growing concerns about the rising tide of extinctions. Biodiversity loss is one of the world's most pressing crises, with many species declining to critically low levels and with significant numbers going extinct. At the same time, there is growing awareness of how biodiversity supports human livelihoods.

CONTENTS:

The Importance of Species

As well as having significant aesthetic, cultural, spiritual and educational values, species form the very foundations of our livelihoods, by providing us with what are known as goods and services. These range from physical goods including food, fuel, clothes and medicine, to essential services such as the purification of water and air, crop pollination and the prevention of soil erosion. Species also provide an invaluable resource for economic activities including fisheries, forestry and tourism.

With the continuing decline of species, nature's ability to provide us with these vital goods and services becomes severely diminished, and the livelihoods of billions of people across the globe are left in jeopardy as a result.

Species and food
The human population is increasing very rapidly, having doubled since 1960, and with this comes a growing need for food.

Species provide us with food both directly and indirectly, and while most people depend primarily on domesticated species for their dietary needs millions of people are dependent on wild species for at least part of their diet.

Since agriculture began about 12,000 years ago, roughly 7,000 plant species have been used for human consumption.

Up to 90% of coastal populations in some parts of the world obtain much of their food and earn their primary income through fishing. Species also contribute to crops; more than 60 wild species have been used to improve the 13 major food crops by providing genes for pest resistance, improved yield, and enhanced nutrition.

The production of at least one third of the world's food, including 87 of the 113 leading food crops, is dependent upon pollination carried out by insects, as well as vertebrates such as bats and birds. This ecosystem service provided by species is worth over US$ 200 billion per year, not including their contribution to the production of crop seeds, forage and pasture grasses, or to maintaining the structure and function of natural ecosystems, which has a value beyond calculation. Without this natural service, not only would those who rely on crop production as a source of income suffer, but millions more would be left without food.

Tomato plants require 'buzz pollination' to reproduce; in order for the anther to release pollen, it must be vibrated at a frequency of around 400 Hz. Hand-held artificial buzzers were used in commercial tomato production, until it was discovered that bumblebees, natural pollinators, buzz at the required frequency. As such, the commercial production of bumblebees for use in greenhouses began in the 1980s, and within three years of the bumblebee, Bombus terrestris, being reared commercially for such use, 95% of all tomato plants in Holland were being pollinated by this species.

Wild species are also important in pest regulation. Bats, toads, birds, snakes, and so forth consume vast numbers of the major animal pests on crops or forests. For example, a single colony of Mexican Free-tailed Bat eats more than 9,000 kilogrammes of insects per night, targeting especially Corn Earthworms and Fall Armyworms, both major crop predators. A single brood of woodpeckers can eat 8,000-12,000 harmful insect pupae per day, helping to maintain the health of forests, whilst in fruit plantations insectivorous birds can make the difference between a bumper crop or a costly failure.

Species and human health

Species are vital to human health for a variety of reasons. In some countries, medicinal plants and animals provide the majority of the drugs people use, whilst freshwater ecosystems, and the species within them, enable the provision and storage of clean water for human use.

Half of the 100 most-prescribed medicinal drugs originate from wild species.

Molluscs, such as clams and mussels, are key contributors in the process of water filtration; a single freshwater bivalve may filter more than seven litres of water a day. Without keystone species such as these, the quality of water in river systems would most likely decline. Dragonflies also play an important role in freshwater ecosystems, by acting as indicators of water quality. If pollution becomes a problem in an area, the dragonflies will be the first to be affected,

and so a reduction in their numbers could indicate a reduction in water quality. This early warning mechanism could prove vital, as action can then be taken to resolve the problem before other species become affected.

More than 70,000 different plant species are used in traditional and modern medicine.

Waste regulation is another important function carried out by species; dung beetles have a dirty job, but their recycling of the droppings of other animals also helps to disperse seeds, recycle nutrients, and suppress parasites. Other species involved in waste regulation include the bottom feeders which survive on dead whales that fall to the bottom of the sea, hyenas and vultures which feed on dead animals, and microbial species which decompose dead vegetation and convert it to fertile soil. Even the seemingly lowliest of species provide critical regulating functions in ecosystems.

Other services provided by species

The importance of species is not limited to the provision of food and medicine; species also carry out what are known as supporting services, including nutrient cycling, photosynthesis and water cycling. Soil formation is also utterly dependent on the species that live in the soil, such as earthworms, termites, and thousands of other invertebrates and microbes. Species-rich soils help to reduce soil erosion, store more carbon, and cycle more water, thereby providing an invaluable contribution to the maintenance of a healthy environment.

Species also hold a cultural value, often acting as key players in our religious, spiritual, psychological and social well-being. They are frequently the subjects of recreational and ecotourism activities including bird-watching and wildlife safaris. The thriving diving industry is fundamentally based on the species of coral, fish and other marine wildlife that people see.

Species are the building blocks of biodiversity and ecosystems and thus the loss of species diminishes the quality of our lives and our basic economic security. However, to save species, we need knowledge so that conservation efforts can be focused to achieve the most effective results. The IUCN Red List of Threatened Species™ provides this essential information on the state of, and trends in, wild species, helps to identify the drivers of biodiversity loss, and identifies where these losses are taking place.

LAMPRODERMA ACANTHOSPORUM KOWALSKI © ALAIN MICHAUD

GREEN TURTLE, CHELONIA MYDAS. SEE PAGE 107 © SHUTTERSTOCK – CIGDEM SEAN COOPER

FLAMBOYANT, DELONIX REGIA. © SHUTTERSTOCK – RUDY UMANS

CHINESE CRESTED TERN, STERNA BERNSTEINI. SEE PAGE 352. © PEI-WEN CHANG

IUCN Red List History

People have an inherent fascination for scarce plants, fungi and animals, and have as a result been documenting the rarity of species for many centuries. In the 1940s several pioneering works were published on extinct and vanishing mammals and birds. Using information from these publications, the early beginnings of the IUCN Red List started in the 1950s with a card index system that was used to document data on threatened mammals and birds. In the early 1960s the card index was transformed into a two-volume set of data sheets. They were presented in loose-leaf format within red binders, with the idea that the data sheets could be replaced when new information became available. The drafts were not available for general circulation, and fewer than fifty were produced.

The first publication of the Red Data Books occurred in 1966. They were titled Red

Data Books with volume 1 on mammals (by Noel Simon) and volume 2 on birds (by James Fisher and Jack Vincent). Volume 3 on reptiles and amphibians (by René Honegger) was published in 1968 and volume 5 on flowering plants (by Ronald Melville) in 1970. Volume 4 on freshwater fishes (by Robert Miller) did not appear until 1977. These were all published in loose-leaf format, so that the data sheets could easily be replaced with updated versions. Starting in the 1970s, a further series of much more detailed IUCN Red Data Books was produced covering mammals, birds, reptiles, plants, invertebrates and swallowtail butterflies. However, as the production costs of these grew, the first IUCN Red List of Threatened Animals was produced in book form in 1986, covering many more species, but with only very limited detail. Updates of the animal Red List were published in 1988, 1990, 1994, and 1996 and an IUCN Red List of Threatened Plants was published in 1997.

The Red Data Books produced by IUCN and its partners between 1949 and 1968 used a variety of qualitative terms and categories (referred to as the Red List Categories) to describe the status of species. In 1969 there was an attempt to standardize the approach and four categories were adopted (Endangered, Rare, Depleted and Indeterminate). These were revised again in 1972 to Endangered, Vulnerable, Rare, Out of Danger and Indeterminate; other categories were added in subsequent editions, including Extinct, Insufficiently Known and Commercially Threatened.

As the production of hard copy publications of the IUCN Red List was too resource-intensive, due to the significant increase in the number of species being assessed, it was decided to develop an electronic database which would be made available via the internet. The 2000 IUCN Red List of Threatened Species™ was the first to be released on the internet. The full IUCN Red List database can be viewed at www.iucnredlist.org. This new approach enabled broader access to the information and allowed for more frequent updates at reduced costs.

Since its early beginnings in 1949, the IUCN Red List has come a very long way from a very simple list of 14 mammals and 13 birds, to an increasingly comprehensive and data-rich electronic compendium covering almost 56,000 species (October 2010). The IUCN Red List of Threatened Species™ is now compiled and produced by the IUCN Species Programme working in close partnership with the Species Survival Commission based on contributions from a network of thousands of scientific experts around the world. These include members of the IUCN Species Survival Commission Specialist Groups, Red List Partners (currently BirdLife International, Botanic Gardens Conservation International, Conservation International, NatureServe, Royal Botanic Gardens Kew, Texas A&M University, Sapienza University of Rome, Wildscreen and the Zoological Society of London), and many others including experts from universities, museums, research institutes and non-governmental organizations.

The IUCN Species Survival Commission, the largest of IUCN's six volunteer commissions with a global membership of over 7,000 experts, initiated a review of the Red List Categories in 1989. There was a period of consultation and testing and a new quantitative system of IUCN Red List Categories and Criteria was adopted in 1994.

The use of the new Red List Categories and Criteria to list commercial marine fish in the 1996 IUCN Red List of Threatened Animals prompted much debate. The debate resulted in a review of the efficacy and suitability of the system of Red List Categories and Criteria, especially in relation to commercially managed species like marine fish.

The criteria review process conducted in the late 1990s resulted in the development of a revised system which was adopted in 2001. This revised system is still in use today. A key decision taken at this time was to keep the system stable to enable genuine changes in species status to be detected rather than to have them obscured by constant modifications to the categories and criteria.

In 1998 the Red List Unit was established within the IUCN Species Programme to oversee the compilation and production of the IUCN Red List, and the Species Survival Commission formed the Red List Committee involving the key partners to provide necessary oversight and future strategic development of the Red List.

About the IUCN Red List

The IUCN Red List of Threatened Species™ (or The IUCN Red List) has a long-established history as the world's most comprehensive information source on the global conservation status of animal, fungus and plant species and their links to livelihoods. Although only 3% of the world's described species have been assessed so far, The IUCN Red List provides a useful snapshot of what is happening to species today and highlights the urgent need for conservation. It is the clarion call for fighting the extinction crisis.

The IUCN Red List is based on an objective system for assessing the risk of extinction of a species. There are eight IUCN Red List Categories based on criteria linked to population trend, size and structure, and geographic range. Species listed as Critically Endangered, Endangered or Vulnerable are collectively described as 'threatened'.

The IUCN Red List is not just a register of names and associated threat categories. It is a rich compendium of information on threats, ecological requirements, habitats, and information on conservation actions

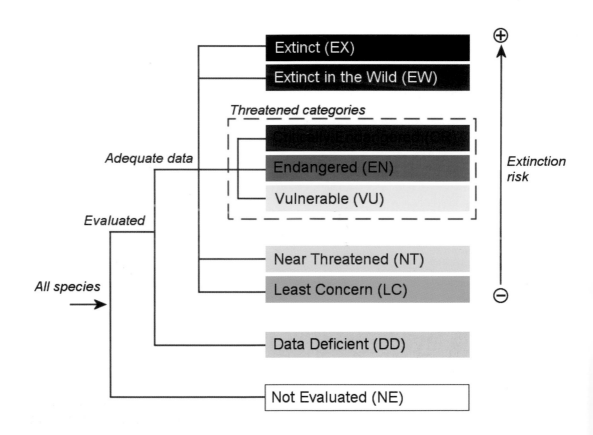

that can be taken to reduce or prevent species extinctions. Species assessments on The IUCN Red List are generated through the knowledge of thousands of the world's leading species scientists through a peer review process.

The 2010 update of The IUCN Red List includes 55,926 species, of which 854 are Extinct or Extinct in the Wild; 18,351 are threatened with extinction (with 3,565 Critically Endangered, 5,256 Endangered and 9,530 Vulnerable); 4,014 are Near Threatened; while 8,358 have insufficient information to determine their threat status (Data Deficient).

Comprehensive assessments of mammal, bird, amphibian, shark, reef-building coral, and cycad and conifer species have been conducted. There are ongoing efforts to complete the assessment of all known species of reptiles, fish, and selected groups of plants, fungi and invertebrates.

Although only a small portion of the world's species has so far been assessed, this sample indicates how life on earth is faring, how little is known, and how urgent the need is to assess more species.

CRITERIA

Population reduction

Restricted geographic range

Small population size & decline

Very small or restricted population

Quantitative analysis

Quantitative thresholds

RED LIST CATEGORIES

Figure 1. Structure of the IUCN Red List Categories and the five IUCN Red List Criteria

Why does the world need The IUCN Red List?

The IUCN Red List provides a benchmark against which to measure the success of conservation actions. Species are easier to identify and categorize than ecosystems, and they are easier to measure than genetic variability. They provide the most useful, and useable, indicators of biodiversity status and loss.

Data from The IUCN Red List can be analyzed and used to:

- identify and document those species most in need of conservation attention
- provide a global index of the state of change of biodiversity

KOMODO DRAGON, VARANUS KOMODOENSIS. SEE PAGE 374 © RICHARD GIBSON

STAGHORN CORAL, ACROPORA CERVICORNIS. SEE PAGE 23 © SHUTTERSTOCK – CECILIA LIM H M

What difference does The IUCN Red List make to species conservation?

One of the aims of The IUCN Red List is to show where conservation action needs to be taken to save the building blocks of our natural capital from extinction. It provides a straightforward way to factor biodiversity needs into decision-making processes, by providing a wealth of useful background information on species. It also acts as a measurement tool to show how trends in the status of species are changing as a result of human activity – positive or negative.

The IUCN Red List is used for:

Conservation and development planning and tools and environmental review
The IUCN Red List is used to identify which areas we need to conserve to protect our most threatened and irreplaceable natural capital. The leading method to determine sites of global significance for biodiversity conservation is application of the Key Biodiversity Area (KBA) guidelines. KBA status is generally determined for sites with threatened species, according to The IUCN Red List. Other criteria not related to The Red List are also used to determine Key Biodiversity Areas. IUCN Red List data is thus essential for identifying the world's most important sites for biodiversity.

For example, The IUCN Red List information on habitat preferences, paired with maps of existing ranges of species, has helped scientists determine where Africa's protected areas and other reserves do not provide adequate protection for the species that need it. Ensuring that protected areas protect biodiversity worldwide is a

EUPHORBIA MILII © SHUTTERSTOCK - SU63

key challenge for the global conservation community during the next decade.

Policy making and influencing legislation
In 2010 the world's governments, including all the member states of the European Union, agreed on a global target as follows:

- By 2020 the extinction of known threatened species has been prevented and their conservation status, particularly of those most in decline, has been improved and sustained.

The IUCN Red List will clearly allow us to monitor progress in relation to this welcome but ambitious target.

IUCN recommends that The IUCN Red List should not be used on its own as a priority-setting tool; in addition to a species' risk of extinction, many other factors must be taken into account, such as costs, logistics, chances of success and other biological characteristics of the species. However, its merit is demonstrated when listings on The IUCN Red List contribute to shaping legislation for species protection. For example, when the Saiga Antelope was listed as Critically Endangered it drew international attention, assisting in achieving agreement to develop a programme for the conservation, restoration and sustainable use of the Saiga Antelope through a Convention of Migratory Species (CMS) agreement.

CRUZIOHYLA CRASPEDOPUS · © SHUTTERSTOCK · DR. MORLEY READ

The IUCN Red List is not only used to influence legislation for individual species. It also contributes to influencing policy and legislation that focuses on area-based conservation. The importance of knowing priority places for conservation, based on species vulnerability as determined by The IUCN Red List, has been growing and gaining legal traction. For example, in the Philippines, an Executive Order was signed in November 2006 mandating the management and protection of Key Biodiversity Areas as critical habitats under the Philippine Wildlife Act.

Motivation for Conservation Action

The 2004 IUCN Red List showed that 30% of amphibians were at risk of extinction. As a result of this report, new conservation groups and alliances were formed to respond to the crisis, resulting in a huge effort to save amphibians from extinction over the past five years.

For example, the IUCN Amphibian Specialist Group was formed and it uses The IUCN Red List categorization to focus attention on conservation projects for Critically Endangered species. Another major international response was the formation of Amphibian Ark. Amphibian Ark is a joint effort of three principal partners: the World Association of Zoos and Aquariums (WAZA), the IUCN SSC Conservation Breeding Specialist Group (CBSG), and the IUCN SSC Amphibian Specialist Group (ASG). This initiative focuses on captive protection for threatened species that would otherwise go extinct if left in the wild. Captive management is one component of the integrated conservation effort which aims to eliminate the threats facing these species in the wild, so that they can be returned to their natural habitats.

In 2009, a new alliance of organizations formed the Amphibian Survival Alliance (ASA), hosted by IUCN, to drive amphibian conservation forward. The priorities for the ASA are the conservation of key habitats for amphibians (since many fall outside of protected areas), and fostering research on key threats, especially on the management of the devastating fungal disease, chytridiomycosis.

Resource allocation
Both governmental and non-governmental organizations increasingly rely on The IUCN Red List to inform priorities, influence legislation, and guide conservation investment. The IUCN Red List has been selected as a key indicator for the Global Environmental Fund (GEF) and is one of the most important data sets used for the new System for Transparent Allocation of Resources (STAR) of the GEF, which determines how much money each country receives. Since its creation in 1991, the GEF has allocated billions of dollars for conservation projects.

Public and private site management
With habitat loss and degradation playing such a large role in the decline of many species, site-based management is an important conservation response. Several companies have begun to use The IUCN Red List data as the baseline for determining actions in operation sites. For example, Holcim (the world's fifth largest cement manufacturer) uses The IUCN Red List to identify the habitat requirements

of species found on their operation sites to assist them in site management and restoration.

Scientific research
Scientific journals of biological sciences and conservation regularly cite The IUCN Red List, and data is downloaded from the website on a daily basis by academics for research purposes. Many leading scientific papers based on analyses of The IUCN Red List have been published in the leading academic journals.

Monitoring trends in biodiversity
The only way to show trends in status for groups of species is through the comparison of sets of IUCN Red List assessments over time. A formal method to determine trends, called The IUCN Red List Index, has been included in various sets of indicators measuring progress towards the United Nations Convention on Biological Diversity's 2020 biodiversity targets.

The proportion of species threatened with extinction is a measure of human impacts on the world's biodiversity, because human activities and their consequences drive the vast majority of threats to biodiversity. Only The IUCN Red List can provide the data to show what proportion of species are threatened globally. A sample of results from several well-known groups provides a glimpse of the severity of the extinction crisis:

- 1 in 4 mammals face extinction
- 1 in 8 birds face extinction
- More than 1 in 3 amphibians face extinction
- 1 in 3 corals face extinction

HALEAKALA SILVERSWORD, ARGYROXIPHIUM SANDWICENSE SSP. MACROCEPHALUM. © SHUTTERSTOCK - AMY TSENG

Awareness-raising, education, communication
Media coverage of reports based on The IUCN Red List extends from printed newspapers to special interest magazines, television and radio coverage. The IUCN Red List is also used by educators and students. Many zoos, aquariums and botanic gardens are using The IUCN Red List information on their animal enclosure signage.

The IUCN Red List website
The IUCN Red List can be viewed in its entirety on The IUCN Red List website

www.iucnredlist.org.

About Species of the Day

The United Nations declared 2010 the International Year of Biodiversity (IYB). This provided a unique opportunity to measure the achievements of governments against the 2010 targets to reduce biodiversity loss and to inspire the world conservation community to set new ambitious conservation targets to save biodiversity, on which all human life depends.

To coincide with the IYB, IUCN launched The IUCN Red List of Threatened Species™ 'Species of the Day' project on 1 January, 2010. The main objective of this initiative was to increase awareness of the huge variety of life and raise the profile of threatened species, in an easy-to-read and visually attractive factsheet which would appeal to a wide-ranging audience. Each species profile provided a photograph, description of the species, key facts, the threats faced, a range map and The IUCN Red List status.

Each day of 2010, a different species was featured, collectively nominated by the majority of the IUCN Species Survival Commission (SSC) network of Specialist Groups. This allowed Species of the Day to represent an entire range of groups – both the charismatic and the obscure species – in all regions, and highlight the range of diverse threats to their existence.

The project was a joint initiative of the IUCN Species Programme and the SSC. It was made possible through the support of UNEP, who kindly provided financial support, and of ARKive, who provided invaluable support both with the writing of the species accounts and sourcing of some of the images. The SSC Specialist Groups, Durrell Wildlife Conservation Trust, BirdLife and the Zoological Society of London also contributed images and text accounts for some of the species profiles.

As well as being displayed on The IUCN Red List website, many other partners of IUCN, organizations, newspapers and interested groups featured Species of the Day on their websites. An enthusiastic and dedicated following also grew from the daily accounts on Twitter, and there was substantial interest in the use of the profiles as an educational tool and for exhibitions on biodiversity.

Species of the Day: Lesser Antillean Iguana

The **Lesser Antillean Iguana**, *Iguana delicatissima*, is listed as 'Endangered' on the IUCN Red List of Threatened Species™. Once common throughout most of the northern Lesser Antilles of the Caribbean, this impressive lizard has been extirpated from several islands and is declining on most others.

Habitat clearance for agriculture was the main historical cause of the decline of this species. Now that tourism has taken over as the region's chief industry, coastal development has led to further habitat loss. Hunting of the Lesser Antillean Iguana was prevalent in the past and, although now illegal, is still common in some areas. Other significant threats include predation by feral mammals such as mongooses, cats and dogs, and hybridization with Common Iguanas (*Iguana iguana*).

The Lesser Antillean Iguana is legally protected throughout its range, but law enforcement is limited. It occurs in several nationally protected areas, but insufficient mitigation of existing threats lessens the conservation impact of these areas. Research on the species' population biology and ecology is ongoing, and captive breeding efforts are underway at the Jersey Wildlife Preservation Trust.

www.iucnredlist.org
www.iucn-iag.org
Help Save Species
www.arkive.org

Species of the Day: Mace Pagoda

The **Mace Pagoda**, *Mimetes stokoei*, is listed as 'Critically Endangered' on the Interim Red Data List of South African Plant Taxa. A member of the protea family, a group of plants characteristic of South Africa's fynbos shrublands, it has a life cycle adapted to fire.

The Mace Pagoda was previously known from a single population in the mountains bordering False Bay in the Western Cape. However, in 1965 an experimental plot was unknowingly established on top of the only existing population (the site was cleared and burned to make way for the new protea orchard); when no more plants were seen after 27 years and two fires, the species was presumed to be extinct. Fortunately, the Mace Pagoda's seeds turned out to remain viable underground for more than 50 years, and in 1999 a runaway wildfire created the very specific conditions needed for germination, producing 24 new seedlings.

The Mace Pagoda is still extremely vulnerable due to its small population size, restricted distribution, and extreme population fluctuations in response to fire. Urgent conservation measures are likely to be needed if the species is to survive.

www.iucnredlist.org
Help Save Species
www.arkive.org

Species of the Day: Livingstone's Flying Fox

Livingstone's Flying Fox, *Pteropus livingstonii*, is one of the largest and most threatened bats in the world and is classified as 'Endangered' on the IUCN Red List of Threatened Species™. Around 1000 individuals remain in the forests clinging onto the precipitous mountain slopes of the western Indian Ocean islands of Anjouan and Mohéli in the Union of the Comoros. Flying Foxes have a vital role in the dynamics of forest ecosystems as they pollinate flowers and disperse seeds.

The Comoros have lost most of their forest and still suffer among the world's highest deforestation rates. The surviving Livingstone's Flying Foxes are highly susceptible to further forest loss, destruction of their habitual roosts and the impacts of natural disasters such as cyclones.

A project recently started to establish the first forest protected areas on these islands to conserve Livingstone's Flying Fox and other wildlife. A captive breeding programme in a number of UK zoos provides a back-up population of this species should the worst happen in the wild.

www.iucnredlist.org
www.durrell.org
Help Save Species

© Alex Sliwa

NOT EVALUATED	DATA DEFICIENT	LEAST CONCERN	NEAR THREATENED	VULNERABLE	‹ ENDANGERED ›	CRITICALLY ENDANGERED	EXTINCT IN THE WILD	EXTINCT
NE	DD	LC	NT	VU	EN	CR	EW	EX

Geographical range

Species of the Day: Tiger

The **Tiger**, *Panthera tigris*, is listed as 'Endangered' on the IUCN Red List of Threatened Species™. The largest of all cats, the Tiger once occurred throughout central, eastern and southern Asia, but currently survives only in scattered populations.

The Caspian, Javan, and Bali Tigers are already extinct, and of the remaining six subspecies, the South China Tiger has not been observed for many years. With approximately 1,400 individuals, India still has the largest national population; however, globally, no more than about 3,200 Tigers roam free in their natural habitat. Poaching and illegal killing are the major threats to the survival of the remaining populations, but habitat loss and overhunting of Tigers and their natural prey species have caused a reduction in distribution, which is now only seven percent of the historic range.

The key to this species' survival is the immediate protection of the remaining populations, and, in the long-term, the maintenance or recovery of large tracts of habitat and corridors, together with the sustainable management of prey populations. This will only be possible through mitigation of the conflicts between local people and Tiger conservation.

www.iucnredlist.org
www.catsg.org
Help Save Species
www.arkive.org

Species of the Day is sponsored by

The production of the IUCN Red List of Threatened Species™ is made possible through the IUCN Red List Partnership: IUCN (including the Species Survival Commission), BirdLife International, Conservation International, NatureServe and Zoological Society of London.

ARUBA ISLAND RATTLESNAKE, CROTALUS UNICOLOR © SHUTTERSTOCK - TIMOTHY CRAIG LUBCKE

This book takes us on a journey, highlighting 365 incredible species that are part of the magnificent biodiversity that surrounds us. The unprecedented loss of biodiversity is a major concern; however, this loss can be slowed down and potentially stopped. A recent study highlighted that 64 mammal, bird and amphibian species have improved in status on The IUCN Red List due to successful targeted conservation action. Furthermore, a detailed investigation has also revealed that an estimated 16 species of birds would probably have gone extinct between 1994-2004 if conservation programmes had not been implemented. Conservation does work and species can recover. However, to halt the extinction crisis urgent action is needed from all of society. It is in everyone's interest to save species, as by saving species, we safeguard biodiversity and the ecosystems that provide the natural resources we need in order to survive.

Dr. Jane Smart OBE
Global Director, IUCN Biodiversity Conservation Group
Director, IUCN Global Species Programme

Dr. Simon Stuart
Chair
IUCN Species Survival Commission

For more information on species, please visit:
www.iucnredlist.org
www.iucn.org/species
www.arkive.org

References

Vie, J.C, Hilton-Taylor, C, Stuart, S. 2008. Wildlife in a Changing World An analysis of the 2008 IUCN Red List of Threatened Species™.

M. Hoffman et el. 2010. The Impact of Conservation on the Status of the World's Vertebrates.

McNeely, J. 2009. The Contribution of Species to Ecosystem Services. IUCN Species Magazine

Hilton-Taylor, C. 2010. A History of The IUCN Red List.

Burton, J.A. 2003. The context of Red Data Books, with a complete bibliography of the IUCN publications. In: H.H. De Long, O.S. Bánki, W. Bergmans and M.J. van der Werff ten Bosch (eds) The Harmonization of Red Lists for Threatened Species in Europe. Proceedings of an International Seminar 27 and 28 November 2002, pp. 291-300. The Netherlands Commission for International Nature Protection, Medelingen No. 38. Leiden.

Coolidge, H.J. 1968. An outline of the origins and growth of the IUCN Survival Service Commission. Transactions of the North American Wildlife and Natural Resources Conference, March 11–13, 1968, 33: 407-417. Wildlife Management Institute, Washington, D.C.

Fitter, R. 1986. Wildlife for Man. How and why we should conserve our species. Collins, London.

Holdgate, M. The Green Web: A Union for World Conservation. Earthscan, London.

Scott, P., Burton, J.A. and Fitter, R. 1987. Red Data Books: the historical background. In: R. Fitter and M. Fitter (eds) The Road to Extinction, pp. 1–6. IUCN, Gland, Switzerland and Cambridge, UK.

Sicilian Fir

Abies nebrodensis

NOT EVALUATED	DATA DEFICIENT	LEAST CONCERN	NEAR THREATENED	VULNERABLE	ENDANGERED	CRITICALLY ENDANGERED	EXTINCT IN THE WILD	EXTINCT
NE	DD	LC	NT	VU	EN	CR	EW	EX

A species is **Critically Endangered** when it faces an extremely high risk of extinction in the wild, based on measurements of population size and/or geographic range and their trends in the past, present and/or future.

© ALJOS FARJON

The **Sicilian Fir**, *Abies nebrodensis*, is listed as 'Critically Endangered' on the IUCN Red List of Threatened Species™. Believed to be extinct at the turn of the 20th Century, the Sicilian Fir was rediscovered in 1957 on the dry slopes of Mount Scalone in north-central Sicily.

Extensive logging and erosion were responsible for the Sicilian Fir's decline.

With only around 30 individuals persisting in the wild today, this flagship species remains perilously close to extinction. Owing to its small population size, the Sicilian Fir is also at inherent risk from any chance event, with fire posing a particular danger.

The remaining wild trees occur within a Regional Park and concerted efforts are being made to ensure its long-term conservation in the wild. In addition, an ex situ conservation programme has been established, producing tens of thousands of trees that will hopefully be successfully replanted in the Sicilian Fir's natural habitat.

Cheetah

Acinonyx jubatus

NOT EVALUATED	DATA DEFICIENT	LEAST CONCERN	NEAR THREATENED	‹ VULNERABLE ›	ENDANGERED	CRITICALLY ENDANGERED	EXTINCT IN THE WILD	EXTINCT
NE	DD	LC	NT	VU	EN	CR	EW	EX

© CHEETAH CONSERVATION FUND

The **Cheetah**, Acinonyx jubatus, is listed as 'Vulnerable' on the IUCN Red List of Threatened Species™. The world's fastest land mammal, the Cheetah once occurred throughout much of Africa and Asia, but has now disappeared from large parts of its former range. Two subspecies, the Northwest African Cheetah and the Asiatic Cheetah, are listed as 'Critically Endangered'.

The main threats to the Cheetah are habitat loss, a reduction in its wild prey, and direct persecution by humans, with Cheetahs often being wrongly perceived as threats to livestock. Competition with other large predators such as Lions can also reduce Cheetah numbers.

This charismatic big cat is legally protected throughout its range and occurs in several reserves, although many are too small to ensure its long-term survival. In some areas, limited trophy-hunting is permitted as an economic incentive to conserve the species. A number of action plans are also in place, and measures such as using guard dogs to protect livestock are helping to reduce the number of Cheetahs being trapped and killed.

A species is Vulnerable when it faces a high risk of extinction in the wild, based on measurements of population size and/or geographic range and their trends in the past, present and/or future.

Atlantic Sturgeon

RED LIST

Acipenser sturio

NOT EVALUATED	DATA DEFICIENT	LEAST CONCERN	NEAR THREATENED	VULNERABLE	ENDANGERED	CRITICALLY ENDANGERED	EXTINCT IN THE WILD	EXTINCT
NE	DD	LC	NT	VU	EN	CR	EW	EX

© Jean-François Hellio and Nicolas Van Ingen (www.hellio-vaningen.fr)

The **Atlantic Sturgeon**, *Acipenser sturio*, is listed as 'Critically Endangered' on the IUCN Red List of Threatened Species™. Sturgeons are one of the oldest fish families in existence, but with 85 percent of all sturgeon species at risk of extinction, they are also the world's most threatened animal group. The Atlantic Sturgeon was previously abundant along all European coasts, but today is restricted to a single, reproductive population that breeds in the Gironde, Garonne and Dordogne basins, France.

Like all sturgeons, this species is long-lived and matures late, increasing its vulnerability to overfishing. It is harvested both for its highly prized flesh and its eggs, which are sold as caviar. Increasing pollution and development, especially channelization, along river systems, has destroyed the spawning and nursery habitats of this species, while the construction of migration obstacles prevents adults from returning to their natal rivers to breed.

International trade in the Atlantic Sturgeon is banned through its listing on Appendix I of the Convention on International Trade in Endangered Species (CITES). A captive breeding programme is being undertaken with the long-term goal of restocking parts of this species' former range.

A species is Critically Endangered when it faces an extremely high risk of extinction in the wild, based on measurements of population size and/or geographic range and their trends in the past, present and/or future.

Staghorn Coral

Acropora cervicornis

NOT EVALUATED	DATA DEFICIENT	LEAST CONCERN	NEAR THREATENED	VULNERABLE	ENDANGERED	CRITICALLY ENDANGERED	EXTINCT IN THE WILD	EXTINCT
NE	DD	LC	NT	VU	EN	CR	EW	EX

© J.E.N. VERON – CORALS OF THE WORLD

The **Staghorn Coral**, Acropora cervicornis, is listed as 'Critically Endangered' on the IUCN Red List of Threatened Species™. With its sister species, A. palmata, it used to be one of the major reef-builders in the Caribbean, and the two are the only species of Staghorn coral found in the Atlantic. There are approximately 160 species of staghorn coral worldwide, the rest of which occur in the Indo-Pacific.

Climate change has a wide range of impacts on corals and the reefs they build, the most important of which are bleaching, acid erosion and increased disease. Climate change also introduces a host of other impacts which may act synergistically with these and local human impacts, including sea level rise, changes to currents, damage from increased storm intensity and frequency, and loss of light from increased river sediment loads.

Staghorn corals are listed on the Convention on the International Trade in Endangered Species (CITES), and also as 'Threatened' on the US Endangered Species Act. For the Caribbean staghorn species, localized efforts to propagate and reintroduce the species have occurred in Florida, Puerto Rico, Dominican Republic, Jamaica and Honduras.

A species is Critically Endangered when it faces an extremely high risk of extinction in the wild, based on measurements of population size and/or geographic range and their trends in the past, present and/or future.

La Palma Stick Grasshopper

Acrostira euphorbiae

NOT EVALUATED	DATA DEFICIENT	LEAST CONCERN	NEAR THREATENED	VULNERABLE	ENDANGERED	CRITICALLY ENDANGERED	EXTINCT IN THE WILD	EXTINCT
NE	DD	LC	NT	VU	EN	CR	EW	EX

© Pedro Oromi

The **La Palma Stick Grasshopper**, *Acrostira euphorbiae*, has not yet been officially evaluated for the IUCN Red List of Threatened Species™, however it is listed as 'Critically Endangered' on the Spanish Red List. It was discovered and described quite recently in 1992, and it is endemic to La Palma, Canary Islands, Spain.

The La Palma Stick Grasshopper is a species that feeds on only one species of plant – *Euphorbia lamarckii*. Due to this high dependence, its occurrence is mainly determined by the distribution of this plant. Recently, the range of the La Palma Stick Grasshopper has been reduced by 30% due to illegal logging of *Euphorbia lamarckii* and a wildfire in 2009. An additional threat is an increase in grazing which leads to both degradation and fragmentation of its habitat. However, the main threat to this species is tourism, with an extensive construction project (hotel and golf course) threatening large parts of the population.

The main measures proposed for the conservation of the La Palma Stick Grasshopper are restoration of the logged and burned areas of its habitat, regulation of grazing intensity, and avoidance of the pressures from tourism.

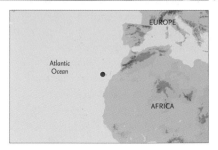

A species is *Critically Endangered* when it faces an extremely high risk of extinction in the wild, based on measurements of population size and/or geographic range and their trends in the past, present and/or future.

Grandidier's Baobab

Adansonia grandidieri

NOT EVALUATED	DATA DEFICIENT	LEAST CONCERN	NEAR THREATENED	VULNERABLE	⟨ ENDANGERED ⟩	CRITICALLY ENDANGERED	EXTINCT IN THE WILD	EXTINCT
NE	DD	LC	NT	VU	EN	CR	EW	EX

Grandidier's Baobab, *Adansonia grandidieri*, is listed as 'Endangered' on the IUCN Red List of Threatened Species™. This large and distinctive tree is found only on Madagascar, where it occurs in scattered locations in the southwest, between Lac Ihotry and Kirindy.

This species is the most heavily exploited of Madagascar's baobabs; the fruits and seeds are collected for food and for the extraction of cooking oil, the bark is used to make ropes, and the spongy wood is dried and sold for thatch. However, the greatest threat to Grandidier's Baobab comes from the transformation of its habitat into agricultural land, in which the species appears to regenerate poorly.

Since 2008, the Global Trees Campaign has been involved in the conservation of this species. Grandidier's Baobab, in particular, is likely to benefit from the creation of protected areas, while formal protection will also be an important step in ensuring its survival. Further research into the extent of bark and fruit exploitation has been recommended, and community education and awareness-raising efforts have been identified as priorities for its conservation.

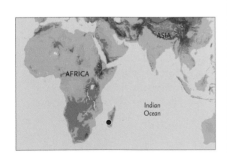

© RICHARD JENKINS

Fony Baobab

Adansonia rubrostipa

NOT EVALUATED	DATA DEFICIENT	LEAST CONCERN	NEAR THREATENED	‹ VULNERABLE ›	ENDANGERED	CRITICALLY ENDANGERED	EXTINCT IN THE WILD	EXTINCT
NE	DD	LC	NT	VU	EN	CR	EW	EX

The **Fony Baobab**, Adansonia rubrostipa, is currently undergoing an update on the IUCN Red List of Threatened Species™, and is likely to be classified as 'Vulnerable'. It occurs in the dry and subarid bioclimates of Madagascar, all along the west coast near Itampolo in the south, to Soalala in the north. This species, the smallest of the six Malagasy baobab species, grows in thicket and dry/spiny forest.

Since the last decade, a considerable amount of the Fony Baobab's habitat has been lost, mainly due to charcoal production, bush fires, and forest clearance for agriculture. Susceptibility to disease is also a problem, and the long generation time of baobabs makes them all the more vulnerable to these threats.

Conservation of the Fony Baobab is currently secured by both its wide distribution and its presence in three protected areas (Namoroka, Kirindy and Tsimanampetsotsa). However, populations north of Toliara are threatened with destruction as a result of charcoal production, and this species will have to be monitored should habitat loss continue at the current rate.

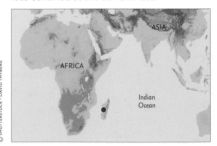

Addax

NOT EVALUATED	DATA DEFICIENT	LEAST CONCERN	NEAR THREATENED	VULNERABLE	ENDANGERED	CRITICALLY ENDANGERED	EXTINCT IN THE WILD	EXTINCT
NE	DD	LC	NT	VU	EN	CR	EW	EX

© Fondation IGF P. Chardonnet

The **Addax**, *Addax nasomaculatus*, is listed as 'Critically Endangered' on the IUCN Red List of Threatened Species™. This desert-dwelling antelope was once found across northern Africa, but is now restricted to a much smaller area, in Niger, Chad, and possibly along the border between Mali and Mauritania.

The Addax has undergone a dramatic decline, mainly as a result of overhunting for its highly prized meat and leather. Other factors have also played a role, including drought and the encroachment of pastoralism into desert lands.

Fewer than 300 Addax now survive in the wild, and its populations are highly fragmented. The largest herd of around 200 animals in the Tin Toumma area of Niger is threatened by oil exploration.

International trade in the Addax is banned under its listing on Appendix I of CITES, and the species is protected under national legislation in Morocco, Tunisia and Algeria. Relatively large numbers of Addax exist in captive populations around the world, and are being used for reintroduction programmes in Tunisia and Morocco.

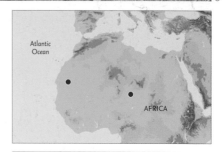

A species is *Critically Endangered* when it faces an extremely high risk of extinction in the wild, based on measurements of population size and/or geographic range and their trends in the past, present and/or future.

Purple Marsh Crab

Afrithelphusa monodosa

NOT EVALUATED	DATA DEFICIENT	LEAST CONCERN	NEAR THREATENED	VULNERABLE	‹ ENDANGERED ›	CRITICALLY ENDANGERED	EXTINCT IN THE WILD	EXTINCT
NE	DD	LC	NT	VU	EN	CR	EW	EX

© Neil Cumberlidge

The **Purple Marsh Crab**, *Afrithelphusa monodosa*, is listed as 'Endangered' on the IUCN Red List of Threatened Species™. This species was first collected in 1947 in Guinea, West Africa, and not seen again until 2005 when a small population was found living in holes in marshy, waterlogged farmland in the Upper Guinea forest in northwest Guinea.

This unusual long-legged purple crab is a semi-terrestrial air-breather, that forages at night and spends the day hiding in waterlogged burrows. It is one of only five rare species of freshwater crabs that belong to an evolutionarily important lineage of freshwater crabs, all of which are endemic to the Upper Guinea rain forests of West Africa.

The continued survival of the Purple Marsh Crab requires the protection of patches of swamp and year-round wetland habitat that are increasingly being converted into farmland. Unfortunately, habitat disturbance and deforestation for agriculture in Guinea and elsewhere in the Upper Guinea forest, and the fact that this species is not found in a protected area, raises questions about the long-term survival of this species.

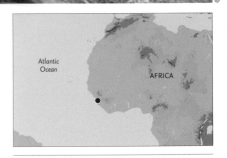

A species is **Endangered** when it faces a very high risk of extinction in the wild, based on measurements of population size and/or geographic range and their trends in the past, present and/or future.

Giant Bronze Gecko

Ailuronyx trachygaster

NOT EVALUATED	DATA DEFICIENT	LEAST CONCERN	NEAR THREATENED	⟨ VULNERABLE ⟩	ENDANGERED	CRITICALLY ENDANGERED	EXTINCT IN THE WILD	EXTINCT
NE	DD	LC	NT	VU	EN	CR	EW	EX

The **Giant Bronze Gecko**, *Ailuronyx trachygaster*, is listed as 'Vulnerable' on the IUCN Red List of Threatened Species™. Found only on the Seychelles islands of Silhouette and Praslin, the Giant Bronze Gecko's sandy-bronze colouration provides it with good camouflage against the branches of its tree habitat.

Although not thought to be currently undergoing a decline, this species is vulnerable to extinction due to its small and restricted distribution. Any habitat degradation within its limited range, for example due to the plausible spread of invasive species, could have disastrous consequences for the survival of the Giant Bronze Gecko.

The Giant Bronze Gecko and its habitat are protected within Praslin National Park, and efforts to control invasive plants are also underway on Praslin Island. On the island of Silhouette, habitat restoration is being undertaken to help this gecko, and the establishment of a new protected area has been recommended to further benefit the species and its forest habitat.

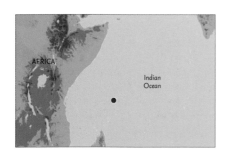

© JUSTIN GERLACH

Giant Panda

Ailuropoda melanoleuca

NOT EVALUATED	DATA DEFICIENT	LEAST CONCERN	NEAR THREATENED	VULNERABLE	‹ ENDANGERED ›	CRITICALLY ENDANGERED	EXTINCT IN THE WILD	EXTINCT
NE	DD	LC	NT	VU	EN	CR	EW	EX

© San Diego Zoo

The **Giant Panda**, *Ailuropoda melanoleuca*, is listed as 'Endangered' on the IUCN Red List of Threatened Species™. Universally admired for its appealing markings and gentle demeanour, this iconic species is a flagship for conservation. Formerly widespread across southern and eastern China, Giant Pandas are now restricted to six mountain ranges at the western edge of their former range.

Habitat restriction and degradation represents the greatest threat to this species. Its range contracted as forests were cleared for agriculture, timber and firewood, leaving small, isolated populations. Giant Pandas are now highly vulnerable to periodic die-backs of bamboo, as they can no longer migrate to find alternative sources of food. Hunting for the fur trade contributed to their historical decline, but this is no longer a significant issue.

Concerted efforts have been made to conserve this species, including the creation of more than 60 panda reserves. The captive breeding programme now includes nearly 300 individuals, reaching the target population size for sustainability. Future plans include the release of captive animals to strengthen wild populations.

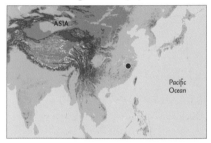

Red Panda

Ailurus fulgens

© RYAN MORGAN

The **Red Panda**, *Ailurus fulgens*, is listed as 'Vulnerable' on the IUCN Red List of Threatened Species™. The taxonomic position of the Red Panda has sparked significant debate, but it is now assigned to a family of its own – Ailuridae. Red Pandas have no close living relatives, and their nearest fossil ancestors, *Parailurus*, lived 3–4 million years ago. This arboreal mammal has a scattered distribution across the Himalayas and mountainous regions of northern Myanmar and southern China.

The principle threat to the Red Panda is the loss of its high-altitude forest habitat, due to logging, agriculture and firewood collection, to meet the demands of the rapidly increasing human population. It may also be hunted for its meat or fur which may be used in traditional dress. Hunting is a particularly significant problem in China.

The Red Panda is protected by law in most range states and occurs in a number of protected areas. However, outside of reserves, hunting of this species continues largely unabated and its habitat is dwindling. Improved protection of its remaining habitat and better law enforcement are crucial for the survival of this species.

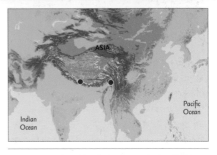

A species is Vulnerable when it faces a high risk of extinction in the wild, based on measurements of population size and/or geographic range and their trends in the past, present and/or future.

Leaf-scaled Sea Snake

NOT EVALUATED	DATA DEFICIENT	LEAST CONCERN	NEAR THREATENED	VULNERABLE	ENDANGERED	CRITICALLY ENDANGERED	EXTINCT IN THE WILD	EXTINCT
NE	DD	LC	NT	VU	EN	CR	EW	EX

© HAROLD COGGER

Although not currently listed on the IUCN Red List of Threatened Species™, the **Leaf-scaled Sea Snake**, *Aipysurus foliosquama*, qualifies as 'Critically Endangered' under IUCN criteria. This reef-associated snake, named for its unusually shaped scales, is known only from Ashmore and Hibernia Reefs, off the northwest coast of Australia.

The Leaf-scaled Sea Snake has undergone a severe population decline over the last 15 years. The causes of this decline are unknown, but may be associated with above-normal sea surface temperatures, which could have sublethal effects on this reef-flat specialist snake. This species may also be impacted by the general degradation of its coral reef habitat. This highly endangered marine reptile occupies a restricted range and probably has restricted dispersal, making it more vulnerable to extinction.

Ashmore Reef is a designated Marine Protected Area, and its protection is actively enforced. However, there are no specific conservation measures currently in place for the Leaf-scaled Sea Snake, and ongoing threats are likely to remain a problem, particularly in light of global climate change.

A species is *Critically Endangered* when it faces an extremely high risk of extinction in the wild, based on measurements of population size and/or geographic range and their trends in the past, present and/or future.

Raso Lark

Alauda razae

NOT EVALUATED	DATA DEFICIENT	LEAST CONCERN	NEAR THREATENED	VULNERABLE	ENDANGERED	CRITICALLY ENDANGERED	EXTINCT IN THE WILD	EXTINCT
NE	DD	LC	NT	VU	EN	CR	EW	EX

The **Raso Lark**, *Alauda razae*, is listed as 'Critically Endangered' on the IUCN Red List of Threatened Species™. With a tiny population confined to a small island in the Cape Verde Islands, the diminutive Raso Lark is perilously close to extinction.

As a result of the species' restricted range, and the frequent periods of drought on Raso Island, its population never rises above around 250 individuals, and following several adverse years without rainfall it may decline to as low as ten breeding pairs. Human disturbance is also a significant threat, and the potential introduction of rats, cats and dogs could be devastating.

Not only is the Raso Lark a protected species under Cape Verde law, but Raso Island requires official authorisation to visit. A priority for the conservation of this species has been to raise awareness within the growing tourist industry to prevent unapproved visits, and in recent years local fishermen have become increasingly committed to helping preserve the island's fragile ecosystem.

© EDWIN WINKEL

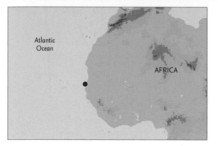

Atlantic Ocean

AFRICA

A species is Critically Endangered when it faces an extremely high risk of extinction in the wild, based on measurements of population size and/or geographic range and their trends in the past, present and/or future.

Chinese Alligator

Alligator sinensis

NOT EVALUATED	DATA DEFICIENT	LEAST CONCERN	NEAR THREATENED	VULNERABLE	ENDANGERED	CRITICALLY ENDANGERED	EXTINCT IN THE WILD	EXTINCT
NE	DD	LC	NT	VU	EN	CR	EW	EX

© GRAHAME WEBB

The **Chinese Alligator**, *Alligator sinensis*, is listed as 'Critically Endangered' on the IUCN Red List of Threatened Species™. One of the smallest and most endangered of the crocodilians, it is restricted to the lower Yangtze valley in the Anhui, Zhejiang and Jiangsu Provinces in China.

Habitat destruction has been the major cause of the Chinese Alligator's decline, and wetland areas continue to be modified for agriculture in an effort to cope with intense human population pressures. There are now estimated to be less than 150 Chinese Alligators remaining in the wild, all of which occur in highly fragmented

subpopulations, each comprising no more than 10 individuals. In contrast to the decimated wild population, the breeding of captive Chinese Alligators has been very successful and the captive population currently exceeds 10,000 individuals.

Chinese authorities have now begun experimental restocking of wild habitats with captive-bred alligators, and are investigating new sites for further reintroductions. The success of the restocking programme is likely to be critical to the long-term survival of the Chinese Alligator.

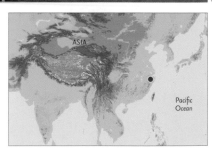

A species is **Critically Endangered** when it faces an extremely high risk of extinction in the wild, based on measurements of population size and/or geographic range and their trends in the past, present and/or future.

Allium pskemense B. Fedtsch

NOT EVALUATED	DATA DEFICIENT	LEAST CONCERN	NEAR THREATENED	VULNERABLE	⟨ ENDANGERED ⟩	CRITICALLY ENDANGERED	EXTINCT IN THE WILD	EXTINCT
NE	DD	LC	NT	VU	EN	CR	EW	EX

© Botanischer Garten TU Darmstadt – Creative Commons Attribution 2.0 Generic (CC BY 2.0)

Allium pskemense B. Fedtsch is listed as 'Endangered' in the Red List Book of Threatened Species of Uzbekistan. Native to central Asia, it is a wild perennial related to the common onion. It is a rare and protected plant, and is restricted in its distribution to the Tien Shan mountains of southern Kazakhstan and Uzbekistan.

Renowned for its curative properties, and due to its widespread use in traditional cooking and medicine, Allium pskemense has been severely overharvested, although there are reports of it being occasionally planted in home gardens.

In Uzbekistan the species is currently being used for research on breeding and, to a lesser extent, for screening for pest and disease resistance (having shown resistance to onion rust *Puccinia allii*). Some demonstration plots of wild Allium have been set up in the Tashkent Botanical Garden to help raise awareness of the importance of conserving crop wild relatives.

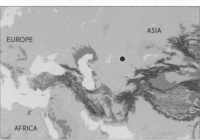

A species is **Endangered** when it faces a very high risk of extinction in the wild, based on measurements of population size and/or geographic range and their trends in the past, present and/or future.

Mazambron Marron

Aloe macra

NOT EVALUATED	DATA DEFICIENT	LEAST CONCERN	NEAR THREATENED	VULNERABLE	‹ ENDANGERED ›	CRITICALLY ENDANGERED	EXTINCT IN THE WILD	EXTINCT
NE	DD	LC	NT	VU	EN	CR	EW	EX

The **Mazambron Marron**, *Aloe macra*, is not currently listed in the IUCN Red List of Threatened Species™, however it has been nationally assessed by the Conservatoire Botanique National de Mascarin as 'Endangered'. This aloe, found only in the dry to semidry habitats of Réunion Island in the Mascarenes, is only known from approximately 200 wild individuals distributed across five sites.

This species has been declining due to the destruction of its habitat, which is among the most damaged of all native vegetation communities on Réunion. Introduced Giant African Land Snails (*Achatina* spp.) are very partial to the plants, and pose a serious additional threat capable of wiping colonies out in little time. The other main threat is predation by rats, which have also been introduced to the island.

Propagation efforts have been highly successful, but attempts to reinforce wild populations have often been met with difficulties owing to the impact of the alien herbivores and predators. The challenge is now to successfully re-establish viable populations in the original habitat, which will depend on the ability to curtail damage by both rats and snails.

© VALÉRIE GRONDIN

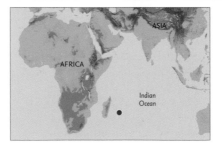

A species is **Endangered** when it faces a very high risk of extinction in the wild, based on measurements of population size and/or geographic range and their trends in the past, present and/or future.

Antiguan Racer

Alsophis antiguae

RED LIST

NOT EVALUATED	DATA DEFICIENT	LEAST CONCERN	NEAR THREATENED	VULNERABLE	ENDANGERED	CRITICALLY ENDANGERED	EXTINCT IN THE WILD	EXTINCT
NE	DD	LC	NT	VU	EN	CR	EW	EX

© Matthew Morton

The **Antiguan Racer**, *Alsophis antiguae*, is listed as 'Critically Endangered' on the IUCN Red List of Threatened Species™. Once abundant across Antigua in the Eastern Caribbean, it was decimated by mongooses, which were introduced in the late 1800s in a futile attempt to control rats in sugarcane plantations. This harmless snake was declared extinct in 1936.

In 1989, 50 individuals were found on Great Bird Island, a nine-hectare islet off the coast of Antigua. The Antiguan Racer Conservation Project was launched to rescue the snake. The racers were threatened by alien rats, which attacked even adult snakes and reduced the numbers of their lizard prey. Rats were eradicated from Great Bird Island and a further 11 offshore islets so that the snakes could start to recover and be reintroduced to new sites.

Today, there are more than 300 Antiguan Racers on four islets. The species remains highly vulnerable to the reinvasion of rats and other alien species, exacerbated by the numerous tourist boats which visit these islets. Local volunteers play a vital role in educating visitors and keeping the islands predator-free.

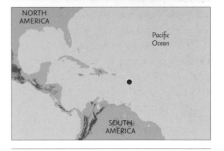

A species is **Critically Endangered** when it faces an extremely high risk of extinction in the wild, based on measurements of population size and/or geographic range and their trends in the past, present and/or future.

Mallorcan Midwife Toad

Alytes muletensis

NOT EVALUATED	DATA DEFICIENT	LEAST CONCERN	NEAR THREATENED	⟨ VULNERABLE ⟩	ENDANGERED	CRITICALLY ENDANGERED	EXTINCT IN THE WILD	EXTINCT
NE	DD	LC	NT	VU	EN	CR	EW	EX

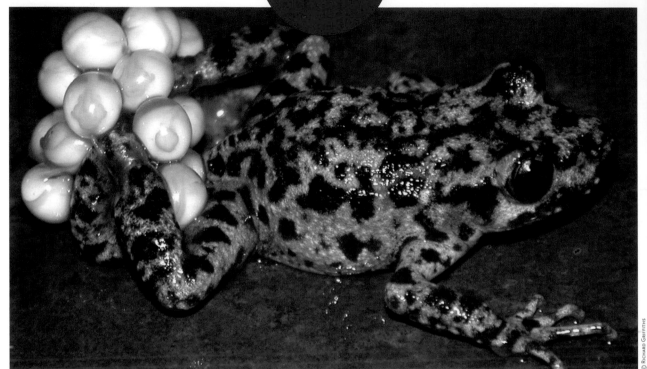

© Richard Griffiths

The **Mallorcan Midwife Toad**, *Alytes muletensis*, is listed as 'Vulnerable' on the IUCN Red List of Threatened Species™. Endemic to the island of Mallorca, this species is also named for its unusual parental care, in which the male carries the developing eggs.

The Mallorcan Midwife Toad is restricted to a mountain range in Mallorca. It has undergone a marked decline as a result of predation by the introduced Viperine Snake and competition with the Perez's Frog, another introduced species. The development of tourism and human settlement on the island is also putting pressure on water resources, and the damming of streams is an additional threat.

As well as being protected under a range of national and international legislation, the Mallorcan Midwife Toad has been the subject of a conservation project involving captive breeding and re-introductions, with several populations already successfully established. Although worries over disease have affected this programme, a new recovery programme is being developed, and control of the Viperine Snake is also underway.

A species is Vulnerable when it faces a high risk of extinction in the wild, based on measurements of population size and/or geographic range and their trends in the past, present and/or future.

Axolotl

NOT EVALUATED	DATA DEFICIENT	LEAST CONCERN	NEAR THREATENED	VULNERABLE	ENDANGERED	CRITICALLY ENDANGERED	EXTINCT IN THE WILD	EXTINCT
NE	DD	LC	NT	VU	EN	CR	EW	EX

© IAN BRIDE

The **Axolotl**, Ambystoma mexicanum, is listed as 'Critically Endangered' on the IUCN Red List of Threatened Species™. This extraordinary species, which persists in a larval stage throughout its life, is restricted to an area of less than 10 square kilometres around Xochimilco on the southern edge of Mexico City.

Whilst there are large numbers of Axolotls in captivity around the world, the wild population is extremely small. The most significant threat to the Axolotl is the increasing pollution of lakes and canals as Mexico City continues to grow. The capture of this species for research and the pet trade contributed to population declines, but the Axolotl now breeds well in captivity thus alleviating this threat.

Conservation efforts for the Axolotl are focusing on raising the profile of Lake Xochimilco through conservation education and a nature tourism initiative, alongside work on habitat restoration. In addition, there are several captive colonies around the world, which may eventually provide opportunities for reintroductions of the Axolotl to parts of its historical range, once the main threats have been addressed.

A species is **Critically Endangered** when it faces an extremely high risk of extinction in the wild, based on measurements of population size and/or geographic range and their trends in the past, present and/or future.

Cainarachi Poison Frog

Ameerega cainarachi

NOT EVALUATED	DATA DEFICIENT	LEAST CONCERN	NEAR THREATENED	⟨ VULNERABLE ⟩	ENDANGERED	CRITICALLY ENDANGERED	EXTINCT IN THE WILD	EXTINCT
NE	DD	LC	NT	VU	EN	CR	EW	EX

RED LIST

© Karl-Heinz Jungfer

The **Cainarachi Poison Frog**, _Ameerega cainarachi_, is listed as 'Vulnerable' on the IUCN Red List of Threatened Species™. This small poison frog is found only in the Cainarachi Valley in the northern part of San Martin Department, Peru, where it is most common in the Huallaga Canyon, between elevations of 250 and 750 metres.

The Cainarachi Poison Frog is vulnerable to the destructive activities within its range. Much of the species' habitat is very close to human settlements, and is therefore under threat from forest conversion to make way for coffee plantations, as well as from firewood collecting and livestock grazing.

Owing to its small range, the survival of the Cainarachi Poison Frog is dependant upon the future management of its habitat. As it is not known from any protected reserves, the most important conservation requirement for this species is the implementation of measures to safeguard the areas in which it is found.

A species is Vulnerable when it faces a high risk of extinction in the wild, based on measurements of population size and/or geographic range and their trends in the past, present and/or future.

Wyoming Toad

Anaxyrus baxteri

NOT EVALUATED	DATA DEFICIENT	LEAST CONCERN	NEAR THREATENED	VULNERABLE	ENDANGERED	CRITICALLY ENDANGERED	EXTINCT IN THE WILD	EXTINCT
NE	DD	LC	NT	VU	EN	CR	EW	EX

The **Wyoming Toad**, *Anaxyrus baxteri*, is listed as 'Extinct in the Wild' on the IUCN Red List of Threatened Species™. As this toad's common name suggests, it is native to Wyoming in the United States. Back in the 1950s it was common and often found in the short grass at the edges of ponds and lakes.

A huge and unexplained decline began in the 1960s, and by the mid-1980s they were considered to be extinct. However, in 1987 a small population was found at Mortenson Lake, and by the 1990s a captive breeding programme using this population was started.

Mortenson Lake is now a National Wildlife Refuge and Wyoming toadlets are annually reintroduced into this habitat. Sadly, these reintroductions have had limited success and it is thought that chytrid fungus, predation, pollution, limited genetic diversity, and a drought-related increase in salinity levels of Mortenson Lake, may have played a role in this. Currently, it appears that the population is continuing to decline.

© SUZANNE L. COLLINS, CNAH

A species is Extinct in the Wild when it is known only to survive in cultivation or in captivity. A species is presumed Extinct in the Wild when exhaustive surveys in known and/or expected habitat, at appropriate times throughout its historic range, have failed to record an individual.

Garrison's Andean Damsel

Andinagrion garrisoni

NOT EVALUATED	DATA DEFICIENT	LEAST CONCERN	NEAR THREATENED	VULNERABLE	ENDANGERED	CRITICALLY ENDANGERED	EXTINCT IN THE WILD	EXTINCT
NE	DD	LC	NT	VU	EN	CR	EW	EX

© ROSSER GARRISON

The **Garrison's Andean Damsel**, *Andinagrion garrisoni*, is listed as 'Near Threatened' on the IUCN Red List of Threatened Species™. This damselfly species is known only from northwestern Argentina, in the southern limits of the South American cloud forest. It adds a colourful note to seepages, pools and still waters at stream edges, from mountain rainforests to highland grasslands.

Although this damselfly is not rare in the area, reduction and fragmentation of its habitat may threaten its future prosperity. Habitat loss in this area is due to increased modification and loss of aquatic environments, caused by selective logging and clear-cutting of the forest for agriculture and petroleum prospecting.

This species has so far not been found within existing protected areas, but surveys are needed to confirm their absence or presence. Further research into its ecology is needed, and monitoring is also required to ensure its existence does not become threatened by habitat reduction in the future.

A species is **Near Threatened** when it does not qualify for Critically Endangered, Endangered or Vulnerable now, but is close to qualifying for or is likely to qualify for a threatened category in the near future.

Chinese Giant Salamander

NOT EVALUATED	DATA DEFICIENT	LEAST CONCERN	NEAR THREATENED	VULNERABLE	ENDANGERED	CRITICALLY ENDANGERED	EXTINCT IN THE WILD	EXTINCT
NE	DD	LC	NT	VU	EN	CR	EW	EX

© MICHAEL LAU

The **Chinese Giant Salamander**, *Andrias davidianus*, is listed as 'Critically Endangered' on the IUCN Red List of Threatened Species™. It is the largest of all amphibian species, sometimes growing up to 1.8 metres in length. The Chinese Giant Salamander was formerly widespread in central, southern and southwestern China.

This once common species has suffered a drastic population decline largely due to overharvesting (it is now considered to be a luxury food). Furthermore, their habitat has suffered from the construction of dams, local pesticides, fertilizers and pollutants. Although this species is commercially farmed, wild individuals still suffer from heavy collecting pressure because of their high value for food and as new stock for breeding farms.

In China, seventeen nature reserves have been established for the conservation of the Chinese Giant Salamander, but there are concerns that development around reserves, and plans for tourism, will disrupt healthy rivers and their habitats. Many Chinese Giant Salamanders are produced in commercial breeding farms yearly and a small number of captive-bred individuals have been released back to the wild.

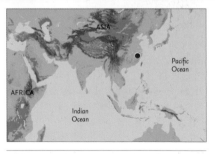

A species is *Critically Endangered* when it faces an extremely high risk of extinction in the wild, based on measurements of population size and/or geographic range and their trends in the past, present and/or future.

Japanese Giant Salamander

Andrias japonicus

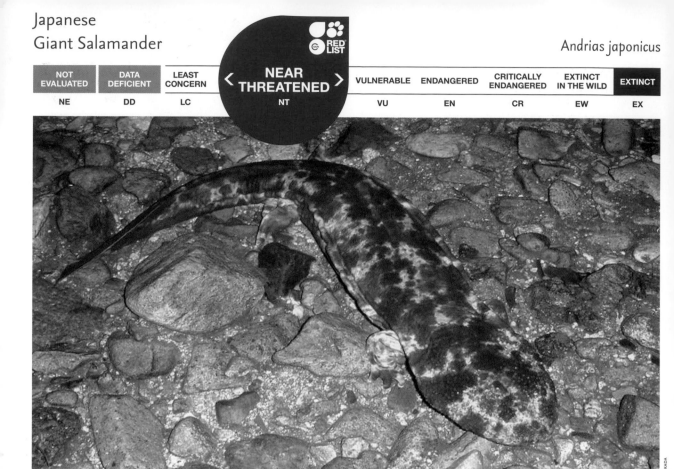

© SUMIO OKADA

The **Japanese Giant Salamander**, *Andrias japonicus*, is listed as 'Near Threatened' on the IUCN Red List of Threatened Species™. Growing to a total length of 150 cm, this is the second largest amphibian in the world, surpassed only by its close relative the Chinese Giant Salamander, *Andrias davidianus*. Endemic to Japan, it is found in small to large rivers in clear, cool, oxygenated water.

This species is threatened mainly by habitat loss and habitat degradation. Whilst populations in remote mountain forest areas are relatively stable, those in lower-lying urbanized regions are highly threatened. Artificial concrete river banks prevent natural reproduction, and dams prevent migration and dispersal. Introductions of Chinese Giant Salamanders since the 1970s further threaten the Japanese Giant Salamanders. Hybrids have been reported in several river systems, although the full extent of the problem has yet to be determined.

Habitat restoration projects (e.g. dam modification and construction of artificial nest sites) have been hugely successful in bringing back urban salamander populations. Community-based conservation initiatives also give this species further hope throughout its range.

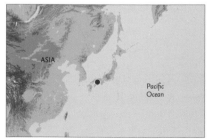

NOT EVALUATED	DATA DEFICIENT	LEAST CONCERN	NEAR THREATENED	VULNERABLE	ENDANGERED	CRITICALLY ENDANGERED	EXTINCT IN THE WILD	EXTINCT
NE	DD	LC	NT	VU	EN	CR	EW	EX

© Lourens Grobler

Angraecum longicalcar is listed as 'Critically Endangered' on the IUCN Red List of Threatened Species™. *Angraecum* is a genus of orchids commonly known as Comet Orchids. *Angraecum longicalcar* has become extinct in some locations (Analavory and Itasy Lake), and is now only found in the rocky Itremo massif in Ambatofinandrahina, Madagascar, where no more than 50 individuals are recorded.

This species has been facing overexploitation mainly for local trade, and habitat loss due to bush fire. Regulated by CITES under Appendix II, it is an offence to trade *Angraecum longicalcar* internationally without a permit. Itremo-Ambatofinandrahina has the potential to become a legally protected area, and management of this site needs to ensure that the only subpopulation of *Angraecum longicalcar* is included within any new protected area delimitation.

Scientific studies of this species, coupled with community-based conservation and reintroduction of seedlings, is being undertaken at the Royal Botanic Gardens, Kew. As the main threat at Itremo-Ambatofinandrahina is bush fire, a fire break is now installed to protect the habitat and fire patrolling will be necessary.

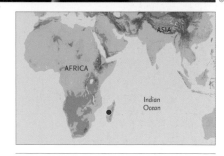

A species is **Critically Endangered** when it faces an extremely high risk of extinction in the wild, based on measurements of population size and/or geographic range and their trends in the past, present and/or future.

European Eel

Anguilla anguilla

NOT EVALUATED	DATA DEFICIENT	LEAST CONCERN	NEAR THREATENED	VULNERABLE	ENDANGERED	CRITICALLY ENDANGERED	EXTINCT IN THE WILD	EXTINCT
NE	DD	LC	NT	VU	EN	CR	EW	EX

© Biopix.dk

The **European Eel**, Anguilla anguilla, is listed as 'Critically Endangered' on the IUCN Red List of Threatened Species™. It spends most of its life in European rivers which flow into the Atlantic, Mediterranean and Baltic seas, and migrates to the western subtropical Atlantic where it breeds.

There is a huge demand in Asia and Europe for European Eels as food and, as a result, they are now being fished unsustainably. Other threats include dams, which have blocked migration routes and cause high mortality rates among migrating eels, and a parasitic nematode introduced from stocked eels from Japan, which is suspected to affect the ability of these eels to reach their spawning grounds.

The International Council for the Exploration of the Sea (ICES) has recommended that a recovery plan be developed as a matter of urgency. International trade in this species is regulated by the Convention on the International Trade in Endangered Species (CITES), and the European Council has issued a regulation requiring the development of eel management plans.

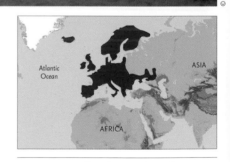

A species is **Critically Endangered** when it faces an extremely high risk of extinction in the wild, based on measurements of population size and/or geographic range and their trends in the past, present and/or future.

NOT EVALUATED	DATA DEFICIENT	LEAST CONCERN	NEAR THREATENED	‹ VULNERABLE ›	ENDANGERED	CRITICALLY ENDANGERED	EXTINCT IN THE WILD	EXTINCT
NE	DD	LC	NT	VU	EN	CR	EW	EX

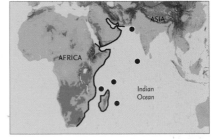

A species is **Vulnerable** when it faces a high risk of extinction in the wild, based on measurements of population size and/or geographic range and their trends in the past, present and/or future.

© DAVID OBURA

Crisp Pillow Coral, Anomastraea irregularis, is listed as 'Vulnerable' on the IUCN Red List of Threatened Species™. This species is found in the western Indian Ocean including the east coast of Africa, the southern Red Sea, Gulf of Aden, Arabian Sea, The Gulf, Madagascar and the Seychelles. The Crisp Pillow Coral is always rare, and unlike many other corals, favours turbid and muddy environments, forming clusters of small colonies.

Because of its rarity, Crisp Pillow Coral is vulnerable to habitat loss, where individual locations or bays may be degraded from activities such as overfishing, pollution and outbreaks of the coral-eating Crown-of-thorns Starfish, *Acanthaster planci*. All corals are threatened by climate change, which is causing ocean acidification and coral bleaching.

Crisp Pillow Coral is currently known from very few Marine Protected Areas (MPAs), but its rarity and aggregation in individual locations makes it amenable to protection through MPAs. Regulations such as a ban on coral collection in Mozambique are helping coral conservation.

It is a candidate species for developing species-specific conservation measures linking new MPAs, fishery management and community conservation and development.

Sahara Aphanius

NOT EVALUATED	DATA DEFICIENT	LEAST CONCERN	NEAR THREATENED	VULNERABLE	ENDANGERED	CRITICALLY ENDANGERED	EXTINCT IN THE WILD	EXTINCT
NE	DD	LC	NT	VU	EN	CR	EW	EX

© HEIKO KAERST

The **Sahara Aphanius**, Aphanius saourensis, is listed as 'Critically Endangered' on the IUCN Red List of Threatened Species™. This small freshwater fish is endemic to the Oued Saoura basin in Algeria. It is now only known from one remnant population, near Mazzer, in the Sahara Desert, having disappeared from numerous other localities within this same spring system.

The North American fish, Gambusia holbrooki, was introduced as a method of mosquito control, but now outnumbers Aphanius by more than 100 to one and poses a serious threat. In addition, excessive groundwater withdrawal for agricultural purposes, the drying of wetlands, and water pollution are also major threats to the species.

Numerous attempts have been made to record Sahara Aphanius from other localities but none have been found. It is unlikely to survive in the wild, however it does well in aquariums, and a small captive breeding programme is underway. This species would benefit from habitat restoration, and the development of a protected area within its range before reintroduction programmes can get underway.

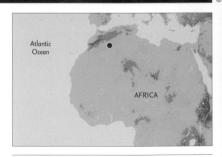

A species is Critically Endangered when it faces an extremely high risk of extinction in the wild, based on measurements of population size and/or geographic range and their trends in the past, present and/or future.

NOT EVALUATED	DATA DEFICIENT	LEAST CONCERN	NEAR THREATENED	VULNERABLE	ENDANGERED	CRITICALLY ENDANGERED	EXTINCT IN THE WILD	EXTINCT
NE	DD	LC	NT	VU	EN	CR	EW	EX

© JUAN RITA LARRUCEA

Apium bermejoi is listed as 'Critically Endangered' on the IUCN Red List of Threatened Species™. This herbaceous plant can be found creeping over the ground in temporary streambeds on the Balearic island of Menorca in the western Mediterranean Sea.

To date, there are less than 100 individuals in an area of just a few dozen square metres. Its habitat is often trampled by passing fishermen and hikers, or more seriously disrupted by offroad motorcyclists. In addition, Apium bermejoi must compete with a wide variety of other plant species for essential water and nutrients. Its present decline seems to be related to a series of drier summers, showing that this species is very sensitive to climate change.

The small area that Apium bermejoi inhabits must be better protected. In 2008, the regional government approved a protection plan, and since then the plant has been reintroduced to three new sites. Seeds have also been collected from each individual to be stored in a seed bank. All populations are monitored at regular intervals; nevertheless, the species continues to be under threat due to its small population size and the possible impacts of climate change.

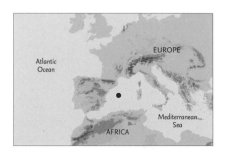

Emperor Penguin

Aptenodytes forsteri

NOT EVALUATED	DATA DEFICIENT	LEAST CONCERN	NEAR THREATENED	VULNERABLE	ENDANGERED	CRITICALLY ENDANGERED	EXTINCT IN THE WILD	EXTINCT
NE	DD	LC	NT	VU	EN	CR	EW	EX

The **Emperor Penguin**, Aptenodytes forsteri, is listed as 'Least Concern' on the IUCN Red List of Threatened Species™. Emperor Penguin breeding colonies occur on areas of stable sea-ice, which may be close to the coast or up to 18 kilometres offshore, and surround the Antarctic continent and nearby islands.

The status of the global population is currently unknown. Only a few colonies have been monitored in detail but, because of its wide-ranging population, the Emperor Penguin is not currently considered to be threatened. However, climate change models predict colossal reductions in the Antarctic sea-ice, and consequently the species breeding habitat. The Emperor Penguin, which appears to be very sensitive to shifts in climate, could therefore decline by as much as 95 percent by 2100. This species may also be negatively affected by increasing ecotourism and by industrial fisheries which deplete its food supply.

Actions are being recommended to reduce disturbance caused by ecotourists. In the face of climate change, the Emperor Penguin would need to adapt, migrate, or change the timing of its breeding season to survive. Therefore, its conservation depends upon climate change mitigation and careful management of its breeding habitat.

© Ty Hurley at The British Antarctic Survey

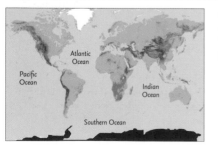

A species is **Least Concern** when it does not qualify for Critically Endangered, Endangered, Vulnerable or Near Threatened. Widespread and abundant species are included in this category.

Nuragica Columbine

Aquilegia nuragica

NOT EVALUATED	DATA DEFICIENT	LEAST CONCERN	NEAR THREATENED	VULNERABLE	ENDANGERED	CRITICALLY ENDANGERED	EXTINCT IN THE WILD	EXTINCT
NE	DD	LC	NT	VU	EN	CR	EW	EX

© GABRIELE CARCANGIU

Exclusively native to one site on the island of Sardinia (Italy), the **Nuragica Columbine**, *Aquilegia nuragica*, is classified as 'Critically Endangered' on the IUCN Red List of Threatened Species™. It is a perennial herb that grows in a gorge along the seasonal Flumineddu River on the nearly vertical limestone cliffs. Nuragica Columbine has only been found on an extremely small site, occupying just a few dozen square metres of a gorge. The number of individuals is very low and they appear to be declining; only 10–15 individuals are believed to exist in this unique population.

Nuragica Columbine is facing a high extinction risk due to natural factors rather than human impact. However, the species is mentioned on several tourist websites, which might attract collectors.

There is no current legal protection in place for Nuragica Columbine, but an action plan needs to be developed and implemented. Collecting interest in this species must also be deterred by removing detailed information concerning the exact location of the species. Cultivating Nuragica Columbine in botanical gardens is recommended.

A species is **Critically Endangered** when it faces an extremely high risk of extinction in the wild, based on measurements of population size and/or geographic range and their trends in the past, present and/or future.

Arabian Tahr

Arabitragus jayakari

NOT EVALUATED	DATA DEFICIENT	LEAST CONCERN	NEAR THREATENED	VULNERABLE	‹ ENDANGERED ›	CRITICALLY ENDANGERED	EXTINCT IN THE WILD	EXTINCT
NE	DD	LC	NT	VU	EN	CR	EW	EX

© David Insall

The **Arabian Tahr**, *Arabitragus jayakari*, is listed as 'Endangered' on the IUCN Red List of Threatened Species™. This relative of the wild goat is confined to the mountains of northern Oman and the United Arab Emirates.

The greatest threat to the survival of the species is the loss and fragmentation of its habitat, but illegal hunting and competition with livestock are also major concerns. In the United Arab Emirates, where the Tahr population is especially fragmented, conservationists fear that the results of inbreeding will increase susceptibility to disease and decrease fertility. In 2009, Wadi Wurayah Fujairah was officially declared as the United Arab Emirates' first protected mountain area.

In Oman, where the largest populations occur, they are well protected in the Wadi Sareen Nature Reserve under the Diwan of Royal Court. The Ministry of Environment and Climate Affair's rangers protect populations in the Sharqiyah and the Western Hajar mountains. Protection of wild populations is essential, because several captive breeding programmes in Oman have shown the Arabian Tahr to be difficult to breed in captivity.

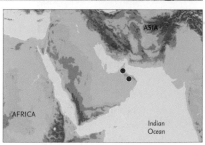

A species is Endangered when it faces a very high risk of extinction in the wild, based on measurements of population size and/or geographic range and their trends in the past, present and/or future.

NOT EVALUATED	DATA DEFICIENT	LEAST CONCERN	NEAR THREATENED	VULNERABLE	ENDANGERED	CRITICALLY ENDANGERED	EXTINCT IN THE WILD	EXTINCT
NE	DD	LC	NT	VU	EN	CR	EW	EX

© GUILLERMO SEIJO, IBONE, VMABCC-BIOVERSITY, 2009

Arachis rigonii is a wild relative of the cultivated peanut and is listed as 'Critically Endangered' in the 'Red List Book of Crop Wild Relatives of Bolivia'. Endemic to Bolivia, its local common name is given as Manicillo rigoni – 'manicillo' indicates the smaller size of its fruit compared to cultivated varieties ('mani' is the Spanish word for peanut), and 'rigoni' refers to the agronomist V.A. Rigoni who first studied it.

Unfortunately, due to habitat fragmentation and loss, this species currently survives as a single, fragmented population in a heavily built-up area in the city of Santa Cruz, occupying an area of less than 10 km² at an altitude of 400 m.

Given its critical conservation status, further research is urgently needed to locate other populations of this species in the area surrounding Santa Cruz. Action needs to be taken to target protection of this species in its natural habitat. Other conservation measures include setting up a seed collection in gene banks and in botanical gardens, as well as establishing herbaria collections (dried and preserved specimens).

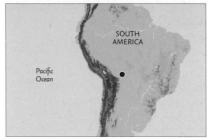

A species is Critically Endangered when it faces an extremely high risk of extinction in the wild, based on measurements of population size and/or geographic range and their trends in the past, present and/or future.

Blue-throated Macaw

Ara glaucogularis

|---|---|---|---|---|---|---|---|---|
| NE | DD | LC | NT | VU | EN | CR | EW | EX |

The **Blue-throated Macaw**, Ara glaucogularis, qualifies as 'Critically Endangered' on the IUCN Red List of Threatened Species™. It is found in a small area of north-central Bolivia known as Los Llanos de Moxos, part of the Beni Savannas.

Live export of Blue-throated Macaws from Bolivia has been banned since 1984, and efforts by conservation organisations have radically reduced this trade. Land clearance for pasture and tree-felling for fuel on private cattle ranches reduces the number of suitable nesting trees, which creates nest-site competition from other species of birds. Indiscriminate hunting to provide feathers for indigenous headdresses may have a small impact on the population in some areas.

Efforts continue to be made to suppress illegal trade in this species and raise awareness through community-based conservation action. Alongside these efforts, continued work on nest guarding and creation of new nesting cavities through the provision of nest boxes are all helping to increase the Blue-throated Macaw's population size.

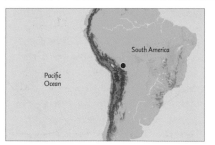

A species is **Critically Endangered** when it faces an extremely high risk of extinction in the wild, based on measurements of population size and/or geographic range and their trends in the past, present and/or future.

54 Blue-throated Macaw • *Ara glaucogularis*

NOT EVALUATED	DATA DEFICIENT	LEAST CONCERN	NEAR THREATENED	VULNERABLE	ENDANGERED	CRITICALLY ENDANGERED	EXTINCT IN THE WILD	EXTINCT
NE	DD	LC	NT	VU	EN	CR	EW	EX

RED LIST

© WOLFGANG KATHE

Arnica montana, commonly referred to as Leopard's Bane, is listed as 'Least Concern' on the IUCN Red List of Threatened Species™. This perennial herb is widespread in Europe, where it is found in extensively managed, acidic and nutrient-poor alpine meadows. Arnica montana flowers are widely used in traditional remedies, for which there is a growing international commercial market, to treat bruises and wounds.

Arnica montana populations are declining in some parts of Europe partly due to unsustainable levels of collection for use in medicines. Its habitat has been maintained in traditional, subsistence-based farming systems, but these are changing due to intensification (especially the addition of nitrogen fertilizers), over-grazing, abandonment, and reforestation. Currently, the largest demand for trade comes from Eastern Europe, the wild resources having been depleted in the main consumer countries – Germany, Switzerland, and the United Kingdom.

Cultivation of Arnica montana has proved to be costly and complicated, and there is a strong market preference for wild-harvested material. The conservation focus is on protocols for non-destructive (sustainable) harvesting and preservation of traditional cultural landscapes. Population trends should be monitored.

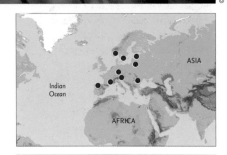

A species is **Least Concern** when it does not qualify for Critically Endangered, Endangered, Vulnerable or Near Threatened. Widespread and abundant species are included in this category.

Long-billed Tailorbird

Artisornis moreaui

NOT EVALUATED	DATA DEFICIENT	LEAST CONCERN	NEAR THREATENED	VULNERABLE	ENDANGERED	CRITICALLY ENDANGERED	EXTINCT IN THE WILD	EXTINCT
NE	DD	LC	NT	VU	EN	CR	EW	EX

The **Long-billed Tailorbird**, *Artisornis moreaui*, is listed as 'Critically Endangered' on the IUCN Red List of Threatened Species™. This small, secretive warbler is found in just two widely separated sites, in the East Usambara Mountains of northeastern Tanzania, and the Njesi Plateau in northern Mozambique.

The Long-billed Tailorbird is under threat from forest destruction and fragmentation, and may have a total population of only a few hundred individuals. The extent of protected forest in the East Usambaras has increased in recent years, but unprotected areas are still under heavy pressure from a range of human activities. The species' preferred gap and edge habitats are also being altered by introduced tree species and by the planting of *Eucalyptus* in areas bordering tea estates.

This species occurs in the Amani Nature Reserve (150–200 individuals) and Nilo Forest Reserve (80 territories) in Tanzania, and various conservation projects are underway, including field surveys, and education and monitoring programmes. The site in Mozambique is relatively undisturbed, but would benefit from official protection, in addition to further survey work.

© Nick Borrow

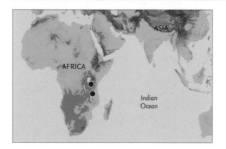

A species is **Critically Endangered** when it faces an extremely high risk of extinction in the wild, based on measurements of population size and/or geographic range and their trends in the past, present and/or future.

NOT EVALUATED	DATA DEFICIENT	LEAST CONCERN	NEAR THREATENED	VULNERABLE	❮ ENDANGERED ❯	CRITICALLY ENDANGERED	EXTINCT IN THE WILD	EXTINCT
NE	DD	LC	NT	VU	EN	CR	EW	EX

© J. ZARUCCHI

Asteropeia micraster is classified as 'Endangered' on the IUCN Red List of Threatened Species™. It grows up to 20 metres in height, in a humid bioclimate, and is native to littoral forests in Madagascar such as those found in Mahabo and the forestry stations of Tampolo and Mandena.

Threats to this species are mainly from humans, since the plant is used for firewood, construction and timber. The species habitat is also under threat due to exploitation of forest timber, mining, fires and a restriction of its range.

The known populations are rare and decreasing in numbers and there are no subpopulations occurring in protected areas. However, a commitment by the Madagascan government to increase the number and size of protected areas in the country has prioritised the protection of littoral forest habitat. Studies have also shown that native littoral species exhibit a dramatic reduction in distribution in response to climate change.

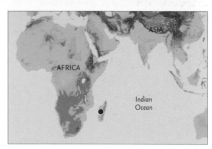

A species is **Endangered** when it faces a very high risk of extinction in the wild, based on measurements of population size and/or geographic range and their trends in the past, present and/or future.

Ploughshare Tortoise

NOT EVALUATED	DATA DEFICIENT	LEAST CONCERN	NEAR THREATENED	VULNERABLE	ENDANGERED	CRITICALLY ENDANGERED	EXTINCT IN THE WILD	EXTINCT
NE	DD	LC	NT	VU	EN	CR	EW	EX

The **Ploughshare Tortoise**, Astrochelys yniphora, is listed as 'Critically Endangered' on the IUCN Red List of Threatened Species™ due to a declining wild population of a few hundred animals. The elongated front spike of the undershell, used by males in breeding jousts for females, is the most remarkable feature of this spectacular tortoise.

Restricted to a tiny area of dry scrubland in northwestern Madagascar, this species has received conservation attention since the early 1970s. Protection of the small population in its natural habitat and a captive breeding programme slowly began to increase its numbers, until it became a target of illegal international wildlife traders.

Though strictly protected under Malagasy and CITES laws, unacceptable numbers of animals are smuggled and sold for huge sums by criminal pet dealers. Enhanced conservation measures are urgently needed to save this species, notably enforcing legal protection to prosecute those who drive the illegal trade, and repatriating recovered animals to secure breeding programmes.

© Anders G.J. Rhodin

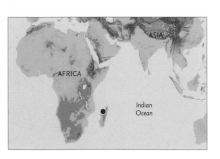

A species is **Critically Endangered** when it faces an **extremely high risk** of extinction in the wild, based on measurements of population size and/or geographic range and their trends in the past, present and/or future.

NOT EVALUATED	DATA DEFICIENT	LEAST CONCERN	NEAR THREATENED	VULNERABLE	ENDANGERED	CRITICALLY ENDANGERED	EXTINCT IN THE WILD	EXTINCT
NE	DD	LC	NT	VU	EN	CR	EW	EX

Although not currently listed on the IUCN Red List of Threatened Species™, Astrophytum caput-medusae qualifies as 'Critically Endangered' based on IUCN assessment criteria. This newly described cactus is named after Medusa, the woman in Greek mythology whose hair was turned to snakes, in reference to this species' long, thin, almost snake-like tubercles.

Astrophytum caput-medusae is endemic to Mexico, where it is known from just a single location, and is undergoing a rapid decline. The species is highly sought-after by collectors, and as a consequence is under threat from illegal collection. In addition, the plants are often trampled by livestock, and its very limited range puts the species at particular risk of extinction.

Very little is currently known about the populations of this newly discovered plant. The species would benefit from effective protection and conservation efforts at the site where it occurs, while off-site propagation measures have also been recommended.

© Ad Konings

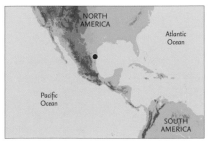

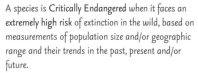
A species is Critically Endangered when it faces an extremely high risk of extinction in the wild, based on measurements of population size and/or geographic range and their trends in the past, present and/or future.

Pale-headed Brush-finch

NOT EVALUATED	DATA DEFICIENT	LEAST CONCERN	NEAR THREATENED	VULNERABLE	ENDANGERED	CRITICALLY ENDANGERED	EXTINCT IN THE WILD	EXTINCT
NE	DD	LC	NT	VU	EN	CR	EW	EX

© Mark Gurney

The **Pale-headed Brush-finch**, Atlapetes pallidiceps, is listed as 'Critically Endangered' on the IUCN Red List of Threatened Species™. Restricted to an extremely small region in southwestern Ecuador, this species was believed to be extinct until its unexpected but thrilling rediscovery in 1998.

Unfortunately, with a current population of just over 100 pairs, and an extremely restricted range, the future of the Pale-headed Brush-finch is still far from certain. Habitat loss from human landscape modification seems to be the primary reason for the bird's limited distribution, while brood parasitism by the Shiny Cowbird, Molothrus bonariensis, is significantly affecting its breeding success.

After the Pale-headed Brush-finch's rediscovery, the small patch of suitable habitat in which it was found, only 27 hectares in size, was purchased to form the Yunguilla Reserve. Intensive management of the species has increased the population in recent years, and is credited with saving the bird from extinction. However, further population growth depends on the acquisition and protection of more habitat, and the continuing control of Shiny Cowbird numbers.

A species is Critically Endangered when it faces an extremely high risk of extinction in the wild, based on measurements of population size and/or geographic range and their trends in the past, present and/or future.

Carossier Palm

NOT EVALUATED	DATA DEFICIENT	LEAST CONCERN	NEAR THREATENED	VULNERABLE	ENDANGERED	CRITICALLY ENDANGERED	EXTINCT IN THE WILD	EXTINCT
NE	DD	LC	NT	VU	EN	CR	EW	EX

The **Carossier Palm**, Attalea crassispatha, is listed as 'Critically Endangered' on the IUCN Red List of Threatened Species™. One of the rarest palms in the Americas, this tall, attractive species is restricted to the southwestern peninsula of Haiti on the island of Hispaniola, in the Caribbean.

The conversion of habitat for agriculture has had a severe impact on the Carossier Palm, and just 30 individuals were estimated to remain in the wild in 1996. Other threats include the harvesting of seeds by people, livestock grazing, and a reduction in the abundance of seed dispersal agents. Given the small size of its population, the Carossier Palm is also vulnerable to extreme natural events.

A large ex situ (outside of its natural habitat) population is maintained at Fairchild Tropical Botanic Garden in Florida. A local NGO, the Fondation Botanique d'Haiti is undertaking a two-year project to investigate the distribution, ecology and conservation status of the Carossier Palm. The project will conduct public-awareness activities, propagate seedlings for outplanting, and carry out further surveys to determine the species' full area of occupancy.

© Scott Zona

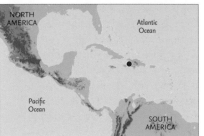

A species is **Critically Endangered** when it faces an extremely high risk of extinction in the wild, based on measurements of population size and/or geographic range and their trends in the past, present and/or future.

Hog Deer

Axis porcinus

NOT EVALUATED	DATA DEFICIENT	LEAST CONCERN	NEAR THREATENED	VULNERABLE	⟨ ENDANGERED ⟩	CRITICALLY ENDANGERED	EXTINCT IN THE WILD	EXTINCT
NE	DD	LC	NT	VU	EN	CR	EW	EX

© Jeremy Holden, Global Wildlife Conservation

The **Hog Deer**, Axis porcinus, is listed as 'Endangered' on the IUCN Red List of Threatened Species™. This small deer once occurred across a large part of Southeast Asia, from Pakistan to southern China, but now has a much more restricted distribution, existing only in highly fragmented, relict populations.

The Hog Deer has undergone a dramatic decline in recent decades, and is now considered one of the most threatened large mammals in parts of its range. The main threats have been hunting, habitat loss and habitat degradation, with the species reported to be easier to hunt than other deer in the region, and its wetland habitats having been largely lost to agriculture and urban development.

The Hog Deer is legally protected throughout its range, and receives protection from international trade under its listing on CITES. It also occurs in a number of protected areas, and various conservation efforts are underway for the species, including education and ecotourism programmes. However, many reserves are too small to allow seasonal movements between habitats, and levels of legal enforcement vary.

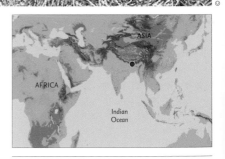

A species is **Endangered** when it faces a very high risk of extinction in the wild, based on measurements of population size and/or geographic range and their trends in the past, present and/or future.

Madagascar Pochard

Aythya innotata

NOT EVALUATED	DATA DEFICIENT	LEAST CONCERN	NEAR THREATENED	VULNERABLE	ENDANGERED	CRITICALLY ENDANGERED	EXTINCT IN THE WILD	EXTINCT
NE	DD	LC	NT	VU	EN	CR	EW	EX

The **Madagascar Pochard**, Aythya innotata, is classified as 'Critically Endangered' on the IUCN Red List of Threatened Species™. It is the rarest duck in the world and, with no sightings for 15 years, it was presumed extinct until 2006 when, surprisingly, 20 individuals were discovered. They were found living in a few remote lakes in the highlands of northwest Madagascar, many hundreds of kilometres from the site where they had last been seen.

Widespread in upland Madagascar in the early twentieth century, the Madagascar Pochard disappeared following the introduction of exotic fish into lakes across the island. Its potential recovery is further threatened by extensive conversion of wetlands to rice production, and certain fishing practices which are deadly to diving ducks.

A species restoration project aims to protect remaining habitat, restore historically occupied wetlands and initiate a captive breeding programme. Eggs collected from nests in October 2009 hatched successfully in captivity and the ducks are now fully grown, thereby doubling the world population and founding a captive population which will hopefully ensure the survival of the Madagascar Pochard.

© IÑAKI RELANZÓN/PHOTOSFERA.COM

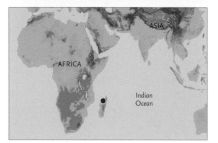

A species is Critically Endangered when it faces an extremely high risk of extinction in the wild, based on measurements of population size and/or geographic range and their trends in the past, present and/or future.

NOT EVALUATED	DATA DEFICIENT	LEAST CONCERN	NEAR THREATENED	VULNERABLE	ENDANGERED	CRITICALLY ENDANGERED	EXTINCT IN THE WILD	EXTINCT
NE	DD	LC	NT	VU	EN	CR	EW	EX

© Robert Ketelaar

The **Socotra Bluet**, *Azuragrion granti*, is currently classified as 'Least Concern' on the IUCN Red List of Threatened Species™, but has recently been reassessed and will be up-listed to 'Near Threatened'. This species is only found on the eastern half of Socotra, Yemen, with its main locality being the Haggeher Mountains, probably due to the higher number of suitable habitats available in that area.

It is found along streams in mountainous habitats, and although this species appears to be relatively common, mounting human pressure in these areas of the island could pose a threat in the future; water resources are progressively being used due to increasing water demands in coastal areas as a result of human population growth and increased tourism.

Due to the very localized occurrence of this species, if climate change were to lead to the island becoming drier as a result of a decrease in precipitation, the Socotra Bluet could quickly be pushed to the edge of extinction.

A species is **Least Concern** when it does not qualify for Critically Endangered, Endangered, Vulnerable or Near Threatened. Widespread and abundant species are included in this category.

NOT EVALUATED	DATA DEFICIENT	LEAST CONCERN	NEAR THREATENED	VULNERABLE	ENDANGERED	CRITICALLY ENDANGERED	EXTINCT IN THE WILD	EXTINCT
NE	DD	LC	NT	VU	EN	CR	EW	EX

RED LIST

© RUPERT KOOPMAN

Babiana blanda is listed as 'Critically Endangered' on the Red List of South African Plants due to its very restricted range and small, declining population. It is distinguished by its large, rosy-pink flowers and striking, deep purplish-blue anthers.

Babiana blanda was believed to be extinct for over 50 years after nearly all the marshy lowland habitat this species inhabits was transformed or degraded by urban and agricultural expansion and invasive alien plants. Fortunately, a small surviving population was discovered in 2006, growing underneath a dense thicket of alien acacias.

The privately-owned site where Babiana blanda was rediscovered also contains 29 other threatened plant species, and negotiations are underway to enable the conservation of this area.

Blue Whale

Balaenoptera musculus

NOT EVALUATED	DATA DEFICIENT	LEAST CONCERN	NEAR THREATENED	VULNERABLE	‹ ENDANGERED ›	CRITICALLY ENDANGERED	EXTINCT IN THE WILD	EXTINCT
NE	DD	LC	NT	VU	EN	CR	EW	EX

© R. HUCKE-GAETE (CBA/UACH)

The **Blue Whale**, *Balaenoptera musculus*, is listed as 'Endangered' on the IUCN Red List of Threatened Species™. Growing to remarkable lengths of around 30 metres, the Blue Whale is the largest animal ever to have lived on the planet. This giant is found in all oceans, ranging from the tropics to the periphery of drift-ice in polar seas, with a preference for open waters.

Before the advent of fast catcher boats and exploding harpoons, the size and speed of the Blue Whale largely protected it from the commercial whaling industry. However, intensive hunting which started in the late 19th century brought the species to the brink of extinction by the 1960s, when it was given international protection. Today, the most significant threat to this species may be the declining availability of krill, its primary food source, whether due to climate change, ocean acidification or other factors.

With hunting of the Blue Whale now prohibited by the International Whaling Commission, there are signs that populations are slowly recovering, although there is still a long way to go before the species is secure.

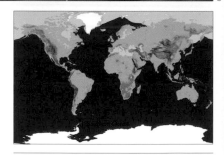

A species is **Endangered** when it faces a very high risk of extinction in the wild, based on measurements of population size and/or geographic range and their trends in the past, present and/or future.

Ethiopian Short-headed Frog

Balebreviceps hillmani

NOT EVALUATED	DATA DEFICIENT	LEAST CONCERN	NEAR THREATENED	VULNERABLE	‹ ENDANGERED ›	CRITICALLY ENDANGERED	EXTINCT IN THE WILD	EXTINCT
NE	DD	LC	NT	VU	EN	CR	EW	EX

The **Ethiopian Short-headed Frog**, *Balebreviceps hillmani*, is listed as 'Endangered' on the IUCN Red List of Threatened Species™. This poorly-known amphibian is endemic to the Bale Mountains of Ethiopia, where it has been recorded from just a single site, at an elevation of around 3,200 metres.

The narrow belt of giant heather woodland inhabited by the Ethiopian Short-headed Frog is under threat by an increasing human population and their livestock. It is not known how damaging this disturbance is to the Ethiopian Short-headed Frog, but future impacts might be disastrous for the species given its potentially very limited distribution. There is also the possibility that the logging of forests at lower elevations may be having indirect negative impacts on its habitat.

The Ethiopian Short-headed Frog receives some protection within the Bale Mountains National Park, although this area has yet to be formally established. High priorities for the conservation of this small amphibian include the effective protection of its habitat, and further surveys to better understand its ecology, status, and the extent of its distribution.

© M. J. LARGEN

A species is **Endangered** when it faces a very high risk of extinction in the wild, based on measurements of population size and/or geographic range and their trends in the past, present and/or future.

Bornean Flat-headed Frog

Barbourula kalimantanensis

NOT EVALUATED	DATA DEFICIENT	LEAST CONCERN	NEAR THREATENED	VULNERABLE	⟨ ENDANGERED ⟩	CRITICALLY ENDANGERED	EXTINCT IN THE WILD	EXTINCT
NE	DD	LC	NT	VU	EN	CR	EW	EX

© David Bickford

The **Bornean Flat-headed Frog**, *Barbourula kalimantanensis*, is listed as 'Endangered' on the IUCN Red List of Threatened Species™. This small frog is found in the Kapuas River Basin in Indonesia. It is the only known frog species to have no lungs; instead it gets all the oxygen it needs through its skin.

As the Bornean Flat-headed Frog is primarily aquatic, found in clear, cold, shallow freshwater rivers, degradation of these habitats has been extremely harmful to its population. Illegal gold-mining and deforestation of the surrounding land has also severely degraded the rivers it inhabits.

Unfortunately, the Bornean Flat-headed Frog is not currently found within any protected areas. Effective conservation of the large remaining forests that surround its range, and better regulation of gold-mining activities, will help preserve its delicate aquatic habitat.

A species is Endangered when it faces a very high risk of extinction in the wild, based on measurements of population size and/or geographic range and their trends in the past, present and/or future.

Myanmar River Turtle

Batagur trivittata

NOT EVALUATED	DATA DEFICIENT	LEAST CONCERN	NEAR THREATENED	VULNERABLE	⟨ ENDANGERED ⟩	CRITICALLY ENDANGERED	EXTINCT IN THE WILD	EXTINCT
NE	DD	LC	NT	VU	EN	CR	EW	EX

© RICK HUDSON

The **Myanmar River Turtle**, *Batagur trivittata*, is listed as 'Endangered' on the IUCN Red List of Threatened Species™, however, a recent re-evaluation of this species suggests that it might become 'Critically Endangered' following completion of the formal Red List review process. Females reach 58 cm in shell length, and mature males develop a spectacular breeding-season colouration, becoming silvery white with black stripes on the shell, with a black face mask over a lime-green head.

Historically, this species occurred throughout the large rivers of Myanmar, with 19th century naturalists reporting huge nesting congregations of basking animals on the delta islands. Intensive egg collection at these predictable sites, capture of adults for consumption, and habitat degradation combined to decimate its populations in the 20th century. Until a few remnant animals were recently found in remote upper tributaries, the status of the species remained uncertain.

Intensive efforts are now underway to work with locals to protect the last known nesting beach, raise juveniles in a safe captive environment for reintroduction, and survey other remote waterways.

Hirola

Beatragus hunteri

The **Hirola**, *Beatragus hunteri*, is listed as 'Critically Endangered' on the IUCN Red List of Threatened Species™. It is confined to a rapidly shrinking corner of northeast Kenya and southern Somalia. There is a small introduced population in Tsavo National Park.

Hirola numbers declined drastically at the end of the 20th century, from 10,000 in the 1970s to just 300 in 1995. The tiny remaining population is now considered to be at risk of imminent extinction. Many of the same factors which led to its rapid decline, such as competition with cattle, severe drought, disease and poaching, present an ongoing threat.

The Hirola Management Committee (HMC) was formed in 1994, with the aim of conserving this species. In 2004, the HMC created the Hirola Strategic Management Plan, which outlined Hirola conservation measures for the next five years. This included creating protected areas, reducing exposure to livestock diseases, monitoring, and promoting income-generating ecotourism for this unique species – measures that will hopefully pull this beautiful antelope back from the edge of extinction.

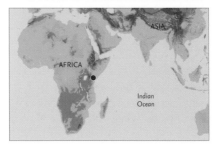

A species is **Critically Endangered** when it faces an **extremely high risk** of extinction in the wild, based on measurements of population size and/or geographic range and their trends in the past, present and/or future.

NOT EVALUATED	DATA DEFICIENT	LEAST CONCERN	NEAR THREATENED	VULNERABLE	ENDANGERED	CRITICALLY ENDANGERED	EXTINCT IN THE WILD	EXTINCT
NE	DD	LC	NT	VU	EN	CR	EW	EX

© Bertrand de Montmollin

Biscutella rotgesii is listed as 'Critically Endangered' on the IUCN Red List of Threatened Species™. This species grows on rocky grassland and serpentine rock in Corsica, tolerating the high concentrations of heavy metals in the soil and surviving on a limited supply of water.

Only several hundred individuals exist, in two populations, and sadly these numbers are diminishing. As Biscutella rotgesii is not internationally protected by any conventions, it is vulnerable to urbanization in its various forms,

particularly the construction of roads, as the main population grows on cliffs surrounding a road. In addition to these direct human pressures, flooding and fires also threaten the existence of this species.

The Conservatoires botaniques nationaux français are storing and cultivating seeds. This important work should ideally be implemented by other botanical gardens. In addition, a management plan for the Natura 2000 site 'Défilé de l'Inzecca' is being drafted and implemented as a way of raising awareness and saving Biscutella rotgesii.

A species is **Critically Endangered** when it faces an extremely high risk of extinction in the wild, based on measurements of population size and/or geographic range and their trends in the past, present and/or future.

European Bison

Bison bonasus

NOT EVALUATED	DATA DEFICIENT	LEAST CONCERN	NEAR THREATENED	‹ VULNERABLE ›	ENDANGERED	CRITICALLY ENDANGERED	EXTINCT IN THE WILD	EXTINCT
NE	DD	LC	NT	VU	EN	CR	EW	EX

© Mieczysław Humecki

The **European Bison**, Bison bonasus, is listed as 'Vulnerable' on the IUCN Red List of Threatened Species™. It is the largest herbivore in Europe and, historically, was distributed throughout western, central and southeastern Europe and the Caucasus.

By the end of the 19th century, the European Bison was close to extinction, with only two wild populations remaining. Habitat degradation and fragmentation due to agricultural activity, forest logging, and unlimited hunting and poaching were the primary reasons for the decrease and extinction of European Bison populations.

As a result of captive breeding, reintroductions, and intensive conservation management, the total population of free-ranging bison has now increased. The captive population is therefore extremely important as a gene reservoir for increasing numbers in the wild. Reintroductions to forests in Belarus, Poland, Russia, Slovakia, Lithuania and the Ukraine have been extremely successful. The saving of the bison has undoubtedly been a success, but further action is essential in order to continue protecting the existing population.

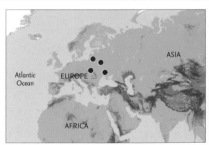

A species is Vulnerable when it faces a high risk of extinction in the wild, based on measurements of population size and/or geographic range and their trends in the past, present and/or future.

Marsh Deer

Blastocerus dichotomus

NOT EVALUATED	DATA DEFICIENT	LEAST CONCERN	NEAR THREATENED	‹ VULNERABLE ›	ENDANGERED	CRITICALLY ENDANGERED	EXTINCT IN THE WILD	EXTINCT
NE	DD	LC	NT	VU	EN	CR	EW	EX

© Patricia Medici

The **Marsh Deer**, Blastocerus dichotomus, is listed as 'Vulnerable' on the IUCN Red List of Threatened Species™. The largest South American deer, it is found in fragmented populations south of the Amazon River, in Argentina, Brazil, Peru, Bolivia and Paraguay, and is considered extinct in Uruguay.

Although the Marsh Deer occurs in several protected areas throughout its range, including in parts of the Pantanal wetlands, it has been lost from much of its former range as a result of conversion of wetland habitat for agriculture, the construction of hydroelectric dams and exotic tree plantations. Cattle ranching has also reduced and fragmented its habitat, and has led to the transmission of 'domestic cattle' diseases to the Marsh Deer. Further threats include poaching and pollution of waterways.

A management plan is urgently needed, with recommended conservation actions including population surveys, ecological studies, improved management of protected areas and the creation of new reserves and protected areas, and the establishment of an international captive breeding programme.

A species is **Vulnerable** when it faces a high risk of extinction in the wild, based on measurements of population size and/or geographic range and their trends in the past, present and/or future.

Salvin's Mushroomtongue Salamander

Bolitoglossa salvinii

RED LIST

NOT EVALUATED	DATA DEFICIENT	LEAST CONCERN	NEAR THREATENED	VULNERABLE	⟨ ENDANGERED ⟩	CRITICALLY ENDANGERED	EXTINCT IN THE WILD	EXTINCT
NE	DD	LC	NT	VU	EN	CR	EW	EX

Salvin's Mushroomtongue Salamander, *Bolitoglossa salvinii*, is listed as 'Endangered' on the IUCN Red List of Threatened Species™. This strikingly-coloured, arboreal salamander occurs in southern Guatemala and in El Salvador. It feeds on small invertebrates, which it catches by shooting out its projectile tongue.

Salvin's Mushroomtongue Salamander has declined as a result of forest loss and fragmentation, mainly due to subsistence agriculture and wood extraction. Although it can persist in shaded coffee and banana plantations and in sugarcane fields, any clearance of these would create habitats too open and dry for its survival. Climate change may also pose a threat by altering moisture conditions, affecting the salamander's ability to take in oxygen through its moist skin.

No specific conservation measures are known to be in place for this intriguing amphibian. However, it potentially occurs in the Parque Nacional El Imposible in El Salvador, and a number of protected areas have been proposed within its range in Guatemala. The maintenance of shaded habitats will be a vital factor in its survival.

© Ted Papenfuss

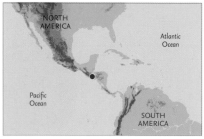

A species is **Endangered** when it faces a very high risk of extinction in the wild, based on measurements of population size and/or geographic range and their trends in the past, present and/or future.

Franklin's Bumble Bee

Bombus franklini

NOT EVALUATED	DATA DEFICIENT	LEAST CONCERN	NEAR THREATENED	VULNERABLE	ENDANGERED	CRITICALLY ENDANGERED	EXTINCT IN THE WILD	EXTINCT
NE	DD	LC	NT	VU	EN	CR	EW	EX

© PETE SCHROEDER

Franklin's Bumble Bee, *Bombus franklini*, is classified as 'Critically Endangered' on the IUCN Red List of Threatened Species™. Known only from southern Oregon and northern California, between the Coast and Sierra-Cascade Ranges in the USA, Franklin's Bumble Bee has the most restricted range of any bumble bee in the world.

Populations of Franklin's Bumble Bee have declined rapidly since 1998, and this species is in imminent danger of extinction. Surveys carried out over more than a decade have illustrated how quickly this bumble bee has disappeared. In 1998, 94 individuals were found at eight sites, whereas in the past four years, only one individual has been observed during surveys.

Threats to this species include: exotic diseases, introduced via trafficking in commercial bumble bees for greenhouse pollination of tomatoes; habitat loss due to destruction, degradation and land conversion; and pesticides and pollution.

A species is **Critically Endangered** when it faces an extremely high risk of extinction in the wild, based on measurements of population size and/or geographic range and their trends in the past, present and/or future.

Banteng

Bos javanicus

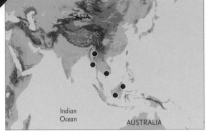

A species is **Endangered** when it faces a very high risk of extinction in the wild, based on measurements of population size and/or geographic range and their trends in the past, present and/or future.

© Brent Huffman / Ultimate Ungulate Images

The **Banteng**, *Bos javanicus*, is listed as 'Endangered' on the IUCN Red List of Threatened Species™. A species of wild cattle, the Banteng occurs in Southeast Asia from Myanmar to Indonesia, with a large introduced population in northern Australia.

The Banteng has been eradicated from much of its historical range, and the remaining wild population, estimated at no more than 8,000 individuals, is continuing to decline. Habitat loss and hunting present the greatest threats to its survival, with the illegal trade in meat and horns still being widespread in Southeast Asia.

Although the Banteng is legally protected across its range and occurs in a number of protected areas, the natural resources of reserves in Southeast Asia often continue to be exploited. The northern Australian population may offer a conservation alternative, although genetic studies hint that the stock may originate from domesticated Bali cattle. Fortunately, a captive population is maintained worldwide which, if managed effectively and supplemented occasionally, can provide a buffer against total extinction, and offer the potential for future reintroductions into the wild.

Dwarf Olive Ibis

NOT EVALUATED	DATA DEFICIENT	LEAST CONCERN	NEAR THREATENED	VULNERABLE	ENDANGERED	CRITICALLY ENDANGERED	EXTINCT IN THE WILD	EXTINCT
NE	DD	LC	NT	VU	EN	CR	EW	EX

The **Dwarf Olive Ibis**, *Bostrychia bocagei*, is listed as 'Critically Endangered' on the IUCN Red List of Threatened Species™. Found only on the island of São Tomé, it may now number fewer than 250 individuals.

The Dwarf Olive Ibis was known only from historical records and anecdotal evidence until its rediscovery in 1990. The clearance of much of São Tomé's forest for sugar cane and cocoa plantations is thought to have had a serious effect on the species, and, although this has now largely ceased, forest is still being cleared for small farms. Predation by introduced mammals such as the Mona Monkey, African Civet and Weasel is also a problem. However, the most serious threat to the Dwarf Olive Ibis is currently hunting by humans.

Measures to protect the Dwarf Olive Ibis's habitat have been proposed, but have yet to be implemented. Research, monitoring programmes and an awareness-raising campaign are underway, but enforced legal protection for the species, and effective protection of remaining forest, will be needed if the species is to survive.

© Nik Borrow

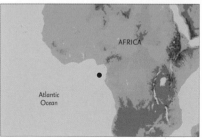

A species is **Critically Endangered** when it faces an extremely high risk of extinction in the wild, based on measurements of population size and/or geographic range and their trends in the past, present and/or future.

Brachyteles
hypoxanthus

NOT EVALUATED	DATA DEFICIENT	LEAST CONCERN	NEAR THREATENED	VULNERABLE	ENDANGERED	CRITICALLY ENDANGERED	EXTINCT IN THE WILD	EXTINCT
NE	DD	LC	NT	VU	EN	CR	EW	EX

© KEVINSCHAFER.COM

The **Northern Muriqui**, Brachyteles hypoxanthus, is listed as 'Critically Endangered' on the IUCN Red List of Threatened Species™. This species occurs in the Atlantic Forest in parts of eastern Brazil, where it has been a flagship species for the conservation of this fragile habitat.

Although once widespread in the Atlantic Forest region, the Northern Muriqui now exists in only a handful of small and isolated subpopulations, none of which are believed to be viable in the long-term. This primate has been widely hunted for food

and sport, and, although hunting is less of a threat today, the widespread destruction of its forests is now putting the Northern Muriqui at serious risk of extinction.

The Northern Muriqui occurs in several protected areas, and a number of research programmes, monitoring programmes and conservation initiatives are in place for the species. In addition, there is a captive breeding programme underway. Effective protection of its remaining habitat will also be vital if this Critically Endangered primate is to survive.

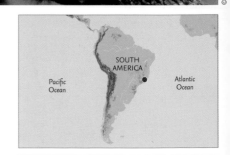

A species is Critically Endangered when it faces an extremely high risk of extinction in the wild, based on measurements of population size and/or geographic range and their trends in the past, present and/or future.

Maned Sloth

Bradypus torquatus

NOT EVALUATED	DATA DEFICIENT	LEAST CONCERN	NEAR THREATENED	VULNERABLE	ENDANGERED	CRITICALLY ENDANGERED	EXTINCT IN THE WILD	EXTINCT
NE	DD	LC	NT	VU	EN	CR	EW	EX

The **Maned Sloth**, *Bradypus torquatus*, is classified as 'Endangered' on the IUCN Red List of Threatened Species™. It is only found in the remaining fragments of the Atlantic Forest of eastern and northeastern Brazil.

This slow-moving, arboreal mammal feeds on the leaves of a very limited number of plants. Its survival is threatened by the continued loss and fragmentation of suitable habitat. Maned Sloths are often killed out of curiosity, but may also be victims of subsistence hunting. There are three genetically distinct isolated populations across the Atlantic Forest, which represent independent targets for conservation efforts.

Halting deforestation is the key to saving the Maned Sloth from extinction. Research on the genetically divergent populations should be promoted to better understand their conservation needs. In view of their low genetic diversity, habitat restoration and corridors are needed to connect the subpopulations within the three main populations.

© Adriano G Chiarello

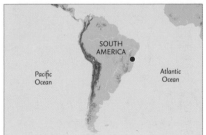

A species is **Endangered** when it faces a very high risk of extinction in the wild, based on measurements of population size and/or geographic range and their trends in the past, present and/or future.

NOT EVALUATED	DATA DEFICIENT	LEAST CONCERN	NEAR THREATENED	VULNERABLE	⟨ ENDANGERED ⟩	CRITICALLY ENDANGERED	EXTINCT IN THE WILD	EXTINCT
NE	DD	LC	NT	VU	EN	CR	EW	EX

Bretschneidera sinensis is listed as 'Endangered' on the IUCN Red List of Threatened Species™. It is the only species in the family Bretschneideraceae, and its distribution extends from central and southern China to Thailand and Vietnam. The population is quite small and its habitat is highly fragmented.

This Chinese tree species is vulnerable to population decline due to the low germination rate of its seeds and poor pollination and plantlet development. In recent decades, natural populations of Bretschneidera sinensis have been greatly reduced by habitat loss and deforestation both for agricultural expansion and for medicinal use. Its attractive flowers also make it a popular ornamental tree, and overcollection of seeds has also contributed to the decline of this species.

Cooperation with local governments and communities is an essential part of the conservation of Bretschneidera sinensis. It has been successfully propagated in many botanical gardens in China, and nature reserves have been established to provide better protection for this species. Bretschneidera sinensis has been successfully reintroduced into the field, and long-term monitoring is being carried out to ensure its survival.

© Bo Lu

A species is **Endangered** when it faces a very high risk of extinction in the wild, based on measurements of population size and/or geographic range and their trends in the past, present and/or future.

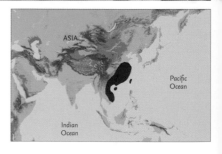

Antsingy Leaf Chameleon

NOT EVALUATED	DATA DEFICIENT	LEAST CONCERN	NEAR THREATENED	⟨ VULNERABLE ⟩	ENDANGERED	CRITICALLY ENDANGERED	EXTINCT IN THE WILD	EXTINCT
NE	DD	LC	NT	VU	EN	CR	EW	EX

The **Antsingy Leaf Chameleon**, *Brookesia perarmata*, is listed as 'Vulnerable' on the IUCN Red List of Threatened Species™. This ornate chameleon species occurs in dry, deciduous forest in Madagascar, where it is only known from the Tsindy de Bemaraha National Park. Although smaller than most chameleons, this species is the largest of the *Brookesia* (dwarf) chameleons.

The Antsingy Leaf Chameleon lives amongst leaf litter and requires primary, relatively untouched forest habitat. Therefore, deforestation caused by expanding agriculture, bush fires and overgrazing threatens this species, especially at the periphery of the national park. Although the Antsingy Leaf Chameleon is listed on Appendix I of CITES, making it illegal to trade this species internationally, illicit collection continues to occur.

The conservation status of this chameleon needs to be updated. A number of other reptiles are endemic to the Tsindy de Bemaraha National Park, and conservation efforts in this area need to be continued. This should be supported by more effective control of the illegal exportation of reptiles from Madagascar's airports and ports.

A species is **Vulnerable** when it faces a high risk of extinction in the wild, based on measurements of population size and/or geographic range and their trends in the past, present and/or future.

Eye of the Crocodile

Bruguiera hainesii

NOT EVALUATED	DATA DEFICIENT	LEAST CONCERN	NEAR THREATENED	VULNERABLE	ENDANGERED	CRITICALLY ENDANGERED	EXTINCT IN THE WILD	EXTINCT
NE	DD	LC	NT	VU	EN	CR	EW	EX

The **Eye of the Crocodile** ('Berus mata buaya' in the Malay language), Bruguiera hainesii, is classified as 'Critically Endangered' on the IUCN Red List of Threatened Species™. This species is found across Singapore, Malaysia, Thailand, Indonesia and Papua New Guinea, with an estimated 200 individuals left in the wild.

The Eye of the Crocodile is a mangrove species that grows in an area known as the 'back mangrove'. This habitat type is at risk of destruction, as it can easily be converted into plantations, agricultural land and shrimp farms.

Several propagation studies have been carried out on the Eye of the Crocodile in an attempt to restore population numbers, but unfortunately it is a slow and frequently unsuccessful process. Further research is needed to fully understand the status of the various populations of this species, and its rate of decline.

© Jean Yong & Jit-Chern Chua

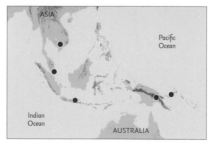

A species is **Critically Endangered** when it faces an extremely high risk of extinction in the wild, based on measurements of population size and/or geographic range and their trends in the past, present and/or future.

Tamaraw

NOT EVALUATED	DATA DEFICIENT	LEAST CONCERN	NEAR THREATENED	VULNERABLE	ENDANGERED	CRITICALLY ENDANGERED	EXTINCT IN THE WILD	EXTINCT
NE	DD	LC	NT	VU	EN	CR	EW	EX

The **Tamaraw**, Bubalus mindorensis, is listed as 'Critically Endangered' on the IUCN Red List of Threatened Species™. It is the largest mammal endemic to the Philippines and can only be found on the island of Mindoro.

The main current threat to the Tamaraw is habitat loss due to farming by resettled and local people. Historically, this species was hunted for both subsistence and sport, which led to a period of drastic decline in numbers of individuals and populations. The introduction of cattle in the past also caused a rinderpest epidemic that contributed to a further decline in numbers. The Tamaraw population has now stabilized and has even shown signs of recovery due to the total ban on sport hunting, closure of nearby ranches, and more intensive patrolling and awareness activities.

The Tamaraw is listed on CITES Appendix I and also receives total protection under Philippine law. A captive breeding programme for this species proved unsuccessful, and so conservation efforts are now focused on protecting the wild population of this charismatic Philippine mammal.

© Josef Suchomel / Ultimate Ungulate Images

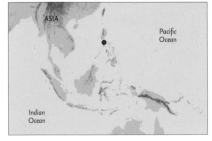

A species is Critically Endangered when it faces an extremely high risk of extinction in the wild, based on measurements of population size and/or geographic range and their trends in the past, present and/or future.

NOT EVALUATED	DATA DEFICIENT	LEAST CONCERN	NEAR THREATENED	VULNERABLE	ENDANGERED	CRITICALLY ENDANGERED	EXTINCT IN THE WILD	EXTINCT
NE	DD	LC	NT	VU	EN	CR	EW	EX

© Jean Bienvenu – west-crete.com

Bupleurum kakiskalae is listed as 'Critically Endangered' on the IUCN Red List of Threatened Species™. The only population, consisting of about 100 individuals, grows on a single limestone cliff in the mountain range of Levka Ori (White Mountains) of western Crete, Greece. This perennial plant grows up to 1 m tall and blossoms with numerous heads of yellow flowers in the late summer.

Given that it flowers only once in its lifetime after 10 to 15 years in a vegetative stage, the Bupleurum kakiskalae struggles to have genetic exchange with other flowering plants. Cliff instability is another threat, as the ground on which it grows periodically collapses. Goats also graze in the area where this plant occurs, feeding indiscriminately on any accessible plant.

As a means of combating Bupleurum kakiskalae's possible extinction, seeds are being collected to conserve the widest possible spectrum of genetic material. In addition, measures are underway to protect it from goats and the threat of collection by overzealous botanists.

A species is **Critically Endangered** when it faces an extremely high risk of extinction in the wild, based on measurements of population size and/or geographic range and their trends in the past, present and/or future.

Mountain Pygmy-possum

Burramys parvus

RED LIST

NOT EVALUATED	DATA DEFICIENT	LEAST CONCERN	NEAR THREATENED	VULNERABLE	ENDANGERED	CRITICALLY ENDANGERED	EXTINCT IN THE WILD	EXTINCT
NE	DD	LC	NT	VU	EN	CR	EW	EX

© Linda Broome

The **Mountain Pygmy-possum**, *Burramys parvus*, is listed as 'Critically Endangered' on the IUCN Red List of Threatened Species™. This is the only Australian mammal that is confined to alpine environments, being found in three isolated populations on mountain summits in southeastern Australia.

The Mountain Pygmy-possum occupies a tiny area, and its populations are highly fragmented. The species' habitat has been affected by introduced mammals, the development of the ski industry in the Australian Alps, and increased temperatures and decreasing snow cover due to climate change. Bushfires and a decline in the species' main prey, the Bogong Moth, are also a problem.

Although the species' entire range occurs within protected areas, important parts of these are in ski resort lease areas. However, management plans are in place for the Mountain Pygmy-possum, and a national recovery plan is being prepared. A range of other recommended conservation measures include habitat restoration, predator control, population and habitat monitoring, and captive breeding programmes.

A species is Critically Endangered when it faces an extremely high risk of extinction in the wild, based on measurements of population size and/or geographic range and their trends in the past, present and/or future.

NOT EVALUATED	DATA DEFICIENT	LEAST CONCERN	NEAR THREATENED	VULNERABLE	ENDANGERED	CRITICALLY ENDANGERED	EXTINCT IN THE WILD	EXTINCT
NE	DD	LC	NT	VU	EN	CR	EW	EX

© Nick Helme

Cadiscus aquaticus is listed as 'Critically Endangered' on the IUCN Red List of Threatened Species™. This succulent green wetland plant, with spectacular white flowers and yellow stamens, is only known from the Western Cape in South Africa, where it grows on shale in a few spring pools. The reproduction strategy of this species is not well known, but the plant is rare and does not spread easily.

Livestock grazing and trampling are likely to have led to the loss of Cadiscus aquaticus in many of its historic localities. Infilling of wetlands and mechanical damage by heavy machinery are considered to be severe ongoing threats. Other less severe threats are invasion by alien grasses caused by the dumping of cattle feed in dry pools during the summer, and eutrophication caused by runoff from fertilizers used on surrounding ploughed lands.

Site management plans and law and policy actions to protect the habitat of Cadiscus aquaticus are urgently needed. More research on the species' ecology is also necessary in order to carry out habitat restoration plans.

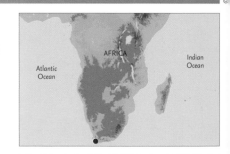

A species is **Critically Endangered** when it faces an extremely high risk of extinction in the wild, based on measurements of population size and/or geographic range and their trends in the past, present and/or future.

Pau Brasil

Caesalpinia echinata

RED LIST

NOT EVALUATED	DATA DEFICIENT	LEAST CONCERN	NEAR THREATENED	VULNERABLE	⟨ ENDANGERED ⟩	CRITICALLY ENDANGERED	EXTINCT IN THE WILD	EXTINCT
NE	DD	LC	NT	VU	EN	CR	EW	EX

The **Pau Brasil**, *Caesalpinia echinata*, is listed as 'Endangered' on the IUCN Red List of Threatened Species™. Famous as the species which gave the country of Brazil its name, the remaining native stands are restricted to the Brazilian states of Pernambuco, Bahia, Espirito Santo and Rio de Janeiro.

In the past, the Pau Brasil was an extremely important source of red dye, a trade that began in the 1500s. The use of synthetic dyes only became widespread in the late 19th century, by which time the natural stands of Brazil's national tree had been all but destroyed. Today, the main threat facing this species is its exportation for the manufacture of bows for stringed instruments.

This flagship species is included on the official list of threatened Brazilian plants, and various laws restrict its export. As part of the Global Trees Campaign, Fauna & Flora International and BGCI have worked to support education and public awareness programmes, and have also collaborated with local partners to carry out further research into the species' distribution and conservation requirements.

© Evan Bowen-Jones

SOUTH AMERICA

Atlantic Ocean

Pacific Ocean

A species is **Endangered** when it faces a very high risk of extinction in the wild, based on measurements of population size and/or geographic range and their trends in the past, present and/or future.

Sea Marigold

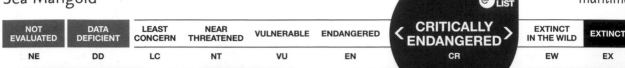

NOT EVALUATED	DATA DEFICIENT	LEAST CONCERN	NEAR THREATENED	VULNERABLE	ENDANGERED	CRITICALLY ENDANGERED	EXTINCT IN THE WILD	EXTINCT
NE	DD	LC	NT	VU	EN	CR	EW	EX

© Angelo Troia

The **Sea Marigold**, *Calendula maritima*, is listed as 'Critically Endangered' on the IUCN Red List of Threatened Species™. This bright yellow flower is found in Sicily and surrounding islets, growing on the decaying remnants of nitrogen-rich sea grass that is washed ashore.

The Sea Marigold is primarily threatened by the encroachment of urban developments on its natural habitat. Since there are currently no legal measures in place to protect this attractive plant, it also faces the threat of collection for its beautiful flowers. In addition to these threats, it has in recent years had to aggressively compete with an alien invasive species, the Iceplant or Hottentot Fig.

For now, it is only the Nature Reserves 'Saline di Trapani e Paceco' and 'Isole dello Stagnone di Marsala' that offer the Sea Marigold protection. Last year, this species was chosen as the official symbol of the province of Trapani. This decision will be accompanied by conservation measures in the field.

A species is **Critically Endangered** when it faces an extremely high risk of extinction in the wild, based on measurements of population size and/or geographic range and their trends in the past, present and/or future.

Red Wolf

NOT EVALUATED	DATA DEFICIENT	LEAST CONCERN	NEAR THREATENED	VULNERABLE	ENDANGERED	CRITICALLY ENDANGERED	EXTINCT IN THE WILD	EXTINCT
NE	DD	LC	NT	VU	EN	CR	EW	EX

The **Red Wolf**, *Canis rufus*, is listed as 'Critically Endangered' on the IUCN Red List of Threatened Species™. One of the world's rarest canids, the Red Wolf formerly ranged throughout southeastern USA, and possibly occurred as far north as Canada. Following a massive decline during the 20th century, the species was declared 'Extinct in the Wild' in 1980.

Historically, the Red Wolf population suffered as a result of persecution and habitat loss. Red Wolves were extensively trapped and shot, as they were believed to pose a direct threat to livestock and game. Today, hybridisation with the closely related coyote (*Canis latrans*) poses the greatest threat to the species.

A highly successful recovery programme has reintroduced the Red Wolf to a 1.7 million-acre area of rural northeastern North Carolina. The wild population is currently doing well, an amazing feat considering the species was at one time 'Extinct in the Wild'. As of March 2010, the free-ranging population numbered an estimated 100–130 individuals, with 78 radio-collared wolves in 29 packs occupying the recovery area.

© Kim Wheeler, Red Wolf Coalition

A species is **Critically Endangered** when it faces an extremely high risk of extinction in the wild, based on measurements of population size and/or geographic range and their trends in the past, present and/or future.

Ethiopian Wolf

Canis simensis

NOT EVALUATED	DATA DEFICIENT	LEAST CONCERN	NEAR THREATENED	VULNERABLE	ENDANGERED	CRITICALLY ENDANGERED	EXTINCT IN THE WILD	EXTINCT
NE	DD	LC	NT	VU	EN	CR	EW	EX

The **Ethiopian Wolf**, *Canis simensis*, is listed as 'Endangered' on the IUCN Red List of Threatened Species™. The wolves are found only in the Highlands of Ethiopia at altitudes above 3,500 m. They live in family packs and prey primarily upon small mammals, many of which are also endemic to the Afroalpine ecosystem.

Less than 500 wolves remain in seven distinct populations, with the majority living in the Bale Mountains National Park. Encroached upon by high altitude agriculture, their Afroalpine habitats continue to shrink, and climate change may push this agriculture frontier even higher.

Pastoralists taking their livestock and their dogs into the wolves' range are responsible for recurrent disease outbreaks of rabies and distemper in the wolf population, which may result in eventual local extinctions.

Conservation measures focus on a mass campaign of dog vaccinations in and around wolves' range to prevent disease transmission. Intensive wolf monitoring, ongoing research, environmental education and protected area support complete a multipronged strategy to secure the long-term survival of this species.

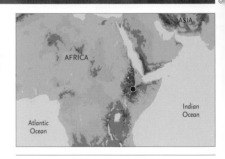

A species is **Endangered** when it faces a very high risk of extinction in the wild, based on measurements of population size and/or geographic range and their trends in the past, present and/or future.

Red Serow

NOT EVALUATED	DATA DEFICIENT	LEAST CONCERN	NEAR THREATENED	VULNERABLE	ENDANGERED	CRITICALLY ENDANGERED	EXTINCT IN THE WILD	EXTINCT
NE	DD	LC	NT	VU	EN	CR	EW	EX

The **Red Serow**, Capricornis rubidus, is listed as 'Near Threatened' on the IUCN Red List of Threatened Species™. The precise distribution of this small, stocky bovid is not known, but it is thought to occur in northern and possibly also western Myanmar, as well as in Assam, India. As its name suggests, this species can be distinguished from other serows mainly by its red coat.

Although little is known about this species, it is believed to be declining as a result of habitat loss and overhunting. Serows are one of the most heavily traded animal species in Myanmar, being hunted for meat and for body parts, which are used in traditional medicine.

International trade in the Red Serow is prohibited under its listing on Appendix I of CITES, and the species may also occur in several protected areas, although this is unconfirmed. However, its legal protection does not appear to be well enforced. Urgent conservation measures include protection from illegal hunting, as well as research into the species' range and populations, and clarification of its taxonomic status.

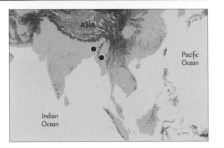

A species is **Near Threatened** when it does not qualify for Critically Endangered, Endangered or Vulnerable now, but is close to qualifying for or is likely to qualify for a threatened category in the near future.

© Gerald Cubitt (geraldcubitt@telkomsa.net)

Puerto Rican Nightjar

NOT EVALUATED	DATA DEFICIENT	LEAST CONCERN	NEAR THREATENED	VULNERABLE	ENDANGERED	CRITICALLY ENDANGERED	EXTINCT IN THE WILD	EXTINCT
NE	DD	LC	NT	VU	EN	CR	EW	EX

© Michael J. Morel

The **Puerto Rican Nightjar**, Caprimulgus noctitherus, is listed as 'Critically Endangered' on the IUCN Red List of Threatened Species™. The Puerto Rican Nightjar is native to southwestern Puerto Rico where there are estimated to be less than 2,000 individuals. Monitoring of this species' population has revealed that this number is continuing to fall drastically.

Habitat loss and degradation have had a significant impact on the population of the Puerto Rican Nightjar, and the problem is ongoing. One potentially serious threat to the species is the proposed wind farm development in Karso del Sur IBA, which could wipe out 5% of the total breeding population. In addition, young birds are at risk of being hunted by Short-eared Owls, and both young and eggs are predated upon by Pearly-eyed Thrashers, fire-ants and feral cats.

The Puerto Rican Nightjar is legally protected but more needs to be done; raising community awareness, effective conservation of existing reserves and continuing research into the species will help to prevent a further decline in numbers.

A species is **Critically Endangered** when it faces an extremely high risk of extinction in the wild, based on measurements of population size and/or geographic range and their trends in the past, present and/or future.

African Golden Cat

Caracal aurata

NOT EVALUATED	DATA DEFICIENT	LEAST CONCERN	NEAR THREATENED	VULNERABLE	ENDANGERED	CRITICALLY ENDANGERED	EXTINCT IN THE WILD	EXTINCT
NE	DD	LC	NT	VU	EN	CR	EW	EX

© Phil Henschel

The **African Golden Cat**, *Caracal aurata*, is listed as 'Near Threatened' on the IUCN Red List of Threatened Species™. It is found in the tropical rainforests of equatorial Africa, where it plays an important role in the traditional beliefs of numerous ethnic groups.

A lack of detailed information on the African Golden Cat makes it difficult to determine its current status in the wild. However, it is believed to have declined due to illegal persecution and the loss and fragmentation of its habitat. This species is frequently caught in illegal cable snares, and skins and whole animals can be readily found in bushmeat markets throughout its range.

The current known range of the African Golden Cat encompasses 18 countries, twelve of which prohibit hunting of the species, whilst two others allow regulated hunting. Larger populations appear to be restricted to well-managed protected areas with suitable forest habitat, although this species may be able to survive to some extent in secondary and degraded forest. Further research is urgently needed to clarify the African Golden Cat's current population status.

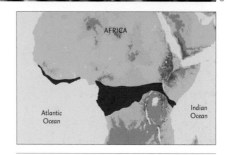

A species is Near Threatened when it does not qualify for Critically Endangered, Endangered or Vulnerable now, but is close to qualifying for or is likely to qualify for a threatened category in the near future.

Oceanic Whitetip Shark

Carcharhinus longimanus

NOT EVALUATED	DATA DEFICIENT	LEAST CONCERN	NEAR THREATENED	‹ VULNERABLE ›	ENDANGERED	CRITICALLY ENDANGERED	EXTINCT IN THE WILD	EXTINCT
NE	DD	LC	NT	VU	EN	CR	EW	EX

© Jeremy Stafford-Deitsch

The **Oceanic Whitetip Shark**, *Carcharhinus longimanus*, is listed as 'Vulnerable' on the IUCN Red List of Threatened Species™. This migratory predator is one of the most widespread shark species, ranging across tropical and subtropical waters in all oceans.

The Oceanic Whitetip Shark suffers from fishing pressure throughout most of its geographic range, with large numbers being caught as by-catch in pelagic fisheries. The shark's large fins are highly prized in international trade, being sold to the Far East to make shark fin soup. Although classified as Vulnerable overall, this species has been assessed as Critically Endangered in the Northwest and Western Central Atlantic, as a result of massive population declines due to by-catch in tuna and billfish fisheries.

Due to the threatening impact of the shark fin trade, steep declines in abundance indices and its highly migratory nature, this species is a suitable candidate for listing under both Appendix II of the Convention on International Trade in Endangered Species (CITES) and Appendix II of the Convention on the Conservation of Migratory Species of Wild Animals.

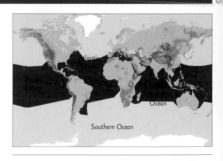

*A species is **Vulnerable** when it faces a high risk of extinction in the wild, based on measurements of population size and/or geographic range and their trends in the past, present and/or future.*

Great White Shark

Carcharodon carcharias

NOT EVALUATED	DATA DEFICIENT	LEAST CONCERN	NEAR THREATENED	‹ VULNERABLE ›	ENDANGERED	CRITICALLY ENDANGERED	EXTINCT IN THE WILD	EXTINCT
NE	DD	LC	NT	VU	EN	CR	EW	EX

© Jeremy Stafford-Deitsch

The **Great White Shark**, *Carcharodon carcharias*, is classified as 'Vulnerable' on the IUCN Red List of Threatened Species™. It can be found almost anywhere in the world, with concentrations in temperate and tropical coastal seas.

The Great White Shark has long been a focus for negative media attention, generated by rare lethal interactions with humans. As a consequence, this species is directly exploited for sports fishing, commercial trophy-hunting, and both the curio and oriental shark-fin trade. Sharks are sold for their flesh, fins, skins, jaws, teeth and oil. Unfortunately, its inquisitive nature and tendency to scavenge from fishing gear, makes this shark vulnerable to either its own accidental entrapment, or deliberate killing by commercial fishermen.

The Great White Shark is currently protected in the Australian EEZ and state waters, South Africa, Namibia, Israel, Malta, Palau and the USA. It should be removed from international game fish record lists, and receive more rational treatment by the media. The recent interest in shark dives and ecotourism may provide a substantial local income and an important method of education.

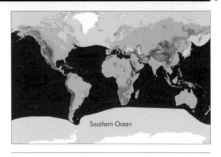

Southern Ocean

A species is **Vulnerable** when it faces a high risk of extinction in the wild, based on measurements of population size and/or geographic range and their trends in the past, present and/or future.

Round Island Keel-scaled Boa

Casarea dussumieri

NOT EVALUATED	DATA DEFICIENT	LEAST CONCERN	NEAR THREATENED	VULNERABLE	‹ ENDANGERED ›	CRITICALLY ENDANGERED	EXTINCT IN THE WILD	EXTINCT
NE	DD	LC	NT	VU	EN	CR	EW	EX

© NIK COLE - DURRELL WILDLIFE CONSERVATION TRUST

The **Round Island Keel-scaled Boa**, *Casarea dussumieri*, is classified as 'Endangered' on the IUCN Red List of Threatened Species™. Following the arrival of people on Mauritius in the 16th century, the boa became restricted to the remote northern island, Round Island. This island has an area of 2.15 km² and currently supports an estimated 1,000 boas. The Round Island Keel-scaled Boa is the only remaining member of the Bolyeridae family, and therefore has a unique jaw adapted to feeding upon other reptiles.

Introduced mammalian predators, such as rats, which decimate island reptile populations, are the main threat to this species. However, Round Island is one of the few islands in the Indian Ocean that has never been invaded by rats. For this reason, the Round Island Keel-scaled Boa still survives.

The removal of introduced rabbits and goats from Round Island, followed by extensive habitat restoration, has allowed reptile, including boa, populations to grow. There are plans to reintroduce this species back within its former range as part of conservation efforts to restore natural island communities around Mauritius.

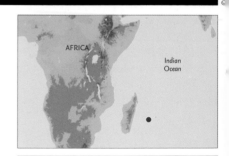

A species is Endangered when it faces a very high risk of extinction in the wild, based on measurements of population size and/or geographic range and their trends in the past, present and/or future.

Chacoan Peccary

Catagonus wagneri

NOT EVALUATED	DATA DEFICIENT	LEAST CONCERN	NEAR THREATENED	VULNERABLE	⟨ ENDANGERED ⟩	CRITICALLY ENDANGERED	EXTINCT IN THE WILD	EXTINCT
NE	DD	LC	NT	VU	EN	CR	EW	EX

The **Chacoan Peccary**, *Catagonus wagneri*, is listed as 'Endangered' on the IUCN Red List of Threatened Species™. The largest peccary species, it is endemic to the dry Chaco of western Paraguay, south-eastern Bolivia and northern Argentina.

The decline in this species is mainly due to hunting for its meat and habitat destruction. All peccary species in the Chaco are heavily hunted, even in protected areas, and the Chacoan Peccary's habitat is being rapidly cleared for agriculture and cattle pasture. In Argentina and Bolivia, much of the land is also overgrazed by livestock and degraded by fire.

The species is legally protected in Argentina, and international trade is prohibited under its listing on Appendix I of CITES. Hunting of all wildlife in Paraguay is also officially prohibited. However, heavy hunting still occurs throughout the species' range, and laws often remain unenforced. Recommended conservation actions for the Chacoan Peccary include upgrading and expanding protected areas, establishing an effective hunting ban, developing environmental education programmes, and expanding captive breeding initiatives.

© Leonardo Maffei

A species is **Endangered** when it faces a very high risk of extinction in the wild, based on measurements of population size and/or geographic range and their trends in the past, present and/or future.

Elegance Coral

Catalaphyllia jardinei

RED LIST

NOT EVALUATED	DATA DEFICIENT	LEAST CONCERN	NEAR THREATENED	⟨ VULNERABLE ⟩	ENDANGERED	CRITICALLY ENDANGERED	EXTINCT IN THE WILD	EXTINCT
NE	DD	LC	NT	VU	EN	CR	EW	EX

Elegance Coral, *Catalaphyllia jardinei*, is listed as 'Vulnerable' on the IUCN Red List of Threatened Species™. This coral is found as individual polyps or in colonies settled loosely in depressions in the sand, from parts of East Africa, to Asia and Australia. While Elegance Coral can catch small prey with its tentacles, it relies upon zooxanthellae, algae living within the coral, to provide it with energy.

This species is rare and is the only member of the *Catalaphyllia* genus. Overharvesting for the aquarium market is the biggest threat to this species and, like all corals, it is also threatened by coral bleaching and ocean acidification, both caused by climate change. Destructive fishing, pollution and coastal development also threaten corals.

Elegance Coral is listed on Appendix II of CITES, which makes it an offence to trade it internationally without a permit. However, quotas on the trade of this species may not be sufficient. It is likely that this coral is protected by some Marine Protected Areas (MPAs), but expansion and creation of MPAs are needed.

A species is Vulnerable when it faces a high risk of extinction in the wild, based on measurements of population size and/or geographic range and their trends in the past, present and/or future.

Kaempfer's Woodpecker

Celeus obrieni

RED LIST

NOT EVALUATED	DATA DEFICIENT	LEAST CONCERN	NEAR THREATENED	VULNERABLE	ENDANGERED	CRITICALLY ENDANGERED	EXTINCT IN THE WILD	EXTINCT
NE	DD	LC	NT	VU	EN	CR	EW	EX

Kaempfer's Woodpecker, *Celeus obrieni*, is listed as 'Critically Endangered' on the IUCN Red List of Threatened Species™. This woodpecker was previously feared extinct before being rediscovered in northeast Tocantins State, central Brazil, in 2006, 80 years after the only previous record. It was long treated as a subspecies of another woodpecker from the Andes, but is distinct in several ways, not least its very different habitat.

The greatest threat to Kaempfer's Woodpecker is the destruction of its Cerrado habitat. Within the region there are huge losses of indigenous Cerrado each year for soya and other crop cultivation, beef production, the cellulose pulp industry and infrastructural developments. The species' habitat is also frequently degraded by criminal arsonism carried out to justify the expansion of cattle ranching.

Recent records suggest that Kaempfer's Woodpecker may be more numerous than currently thought; it is now known to range discontinuously through some 280,000 square kilometres, showing a strong association with *Gadua* bamboo. However, with so little known about this elusive species, there is a pressing need for further surveys to determine the full extent of its range and to estimate its population size.

© CIRO ALBANO

A species is Critically Endangered when it faces an extremely high risk of extinction in the wild, based on measurements of population size and/or geographic range and their trends in the past, present and/or future.

Gunnison Sage-grouse

Centrocercus minimus

A species is **Endangered** when it faces a very high risk of extinction in the wild, based on measurements of population size and/or geographic range and their trends in the past, present and/or future.

The **Gunnison Sage-grouse**, *Centrocercus minimus*, is listed as 'Endangered' on the IUCN Red List of Threatened Species™. This bird, once found in sagebrush habitats in Colorado and Utah, USA, now has a global wild population of fewer than 5,000 and is located in less than 9 percent of its historical range. Only eight populations are known, several of which are estimated to have no more than a hundred individuals.

Potential causes of the species' decline are varied and numerous. Current threats include continued conversion of sagebrush habitat for agricultural purposes, urbanization, and increased recreation, all of which may result in greater predation and habitat loss.

The Gunnison Sage-grouse has been recognized by the American Ornithological Union as one of the ten most endangered bird species in North America. It is therefore listed as a possible candidate for future protection under the Endangered Species Act (ESA). However, while numerous groups have developed conservation plans and actions for the species, it continues to be at significant risk and lacks federal protection under the ESA.

Burrowes' Giant Glass Frog

Centrolene ballux

The **Burrowes' Giant Glass Frog**, *Centrolene ballux*, is listed as 'Critically Endangered' on the IUCN Red List of Threatened Species™. It can be found in humid mountain forests in the Saloya River Valley in Ecuador and the Pacific versant in Colombia. Sadly, the number of Burrowes' Giant Glass Frogs is declining rapidly.

This Giant Glass Frog is primarily threatened by changes to its natural habitat. Climate change is thought to have altered humidity levels in its existing habitat, making it less suitable. In addition, its aquatic habitat is threatened by pollution and the introduction of predatory fish species. Human settlements and activities on forest boundaries are also putting this frog at risk.

This Giant Glass Frog can be found in the privately-owned La Planada reserve in Colombia. More surveys are urgently needed to determine the population status of this species.

© Ignacio De La Riva

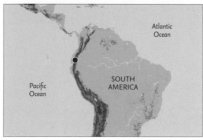

A species is **Critically Endangered** when it faces an extremely high risk of extinction in the wild, based on measurements of population size and/or geographic range and their trends in the past, present and/or future.

Wrinkle-faced Bat

RED LIST

NOT EVALUATED	DATA DEFICIENT	LEAST CONCERN	NEAR THREATENED	VULNERABLE	ENDANGERED	CRITICALLY ENDANGERED	EXTINCT IN THE WILD	EXTINCT
NE	DD	LC	NT	VU	EN	CR	EW	EX

The **Wrinkle-faced Bat**, Centurio senex, is listed as 'Least Concern' on the IUCN Red List of Threatened Species™. This striking-looking bat inhabits moist and dry tropical forests of Latin America, from Mexico to Colombia and Venezuela. It is considered to be an uncommon species throughout its range, although occasionally it can be locally abundant under certain conditions.

It is thought that the species roosts in forest vegetation and not in caves. Males have a fold of skin covered with white fur on the neck that is pulled over the bat's face when resting. As the Wrinkle-faced Bat tends to be found in relatively undisturbed forests, the primary threats to the survival of this species are habitat destruction and deforestation.

The Wrinkle-faced Bat is known from several protected areas in southern Mexico and other countries, but no formal, focused conservation efforts for this species have been launched. Effective protection of tropical forests is required to prevent its further decline.

© Dr. Rodrigo A. Medellín

A species is **Least Concern** when it does not qualify for Critically Endangered, Endangered, Vulnerable or Near Threatened. Widespread and abundant species are included in this category.

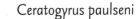

NOT EVALUATED	DATA DEFICIENT	LEAST CONCERN		NEAR THREATENED	VULNERABLE	ENDANGERED	CRITICALLY ENDANGERED	EXTINCT IN THE WILD	EXTINCT
NE	DD	LC		NT	VU	EN	CR	EW	EX

© RICHARD GALLON

The **Baboon Spider**, *Ceratogyrus paulseni*, has not yet been officially evaluated for the IUCN Red List of Threatened Species™, but a provisional assessment of its status is thought to be 'Least Concern'. This species was discovered in 2004 during a survey undertaken in the Kruger National Park in South Africa. It is endemic to South Africa, and found exclusively in a small part of the Letaba area of the park associated with Mopani-Acacia woodland.

The Baboon Spider is a ground-dweller that constructs relatively permanent silk-lined burrows in the ground. They are predominantly sit-and-wait hunters.

Although this species is presently found in a conserved area, its future is threatened by illegal exploitation for national and international trade. Other threats include the effects of predatory species, drought and field fires.

The area where the spiders are found was extensively surveyed over a two-year period and only 20 burrows were located. Baboon Spiders are nationally protected in South Africa, and keeping or selling them without a permit is prohibited. Further research into their ecology, taxonomy and distribution has been recommended to help in their conservation.

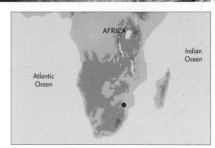

A species is *Least Concern* when it does not qualify for Critically Endangered, Endangered, Vulnerable or Near Threatened. Widespread and abundant species are included in this category.

Basking Shark

NOT EVALUATED	DATA DEFICIENT	LEAST CONCERN	NEAR THREATENED	⟨ VULNERABLE ⟩	ENDANGERED	CRITICALLY ENDANGERED	EXTINCT IN THE WILD	EXTINCT
NE	DD	LC	NT	VU	EN	CR	EW	EX

The **Basking Shark**, *Cetorhinus maximus*, is listed as 'Vulnerable' on the IUCN Red List of Threatened Species™. It gets its name from its habit of 'basking' while surface foraging to filter out its planktonic prey. The Basking Shark is the second largest fish in the world, and is widely distributed in cool temperate waters throughout the world's coastal seas and oceans.

This species was hunted for centuries to supply liver oil for street lighting and industrial use, skin for leather, and flesh for food or fishmeal. Due to its slow reproductive rate, this species is particularly vulnerable to overfishing, and targeted populations are quickly destroyed and are very slow to recover. Today, the biggest threat comes from the demand for fins in the Far East and from accidental by-catch in the fishing industry.

The Basking Shark is now protected in the territorial waters of several countries, and in 2002 it was accepted onto Appendix II of the Convention on International Trade in Endangered Species (CITES), which requires that international trade is monitored.

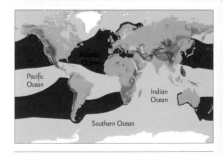

A species is Vulnerable when it faces a high risk of extinction in the wild, based on measurements of population size and/or geographic range and their trends in the past, present and/or future.

Doumergue's Skink

Chalcides parallelus

NOT EVALUATED	DATA DEFICIENT	LEAST CONCERN	NEAR THREATENED	VULNERABLE	⟨ ENDANGERED ⟩	CRITICALLY ENDANGERED	EXTINCT IN THE WILD	EXTINCT
NE	DD	LC	NT	VU	EN	CR	EW	EX

© ROBERT SINDACO

Doumergue's Skink, *Chalcides parallelus*, is listed as 'Endangered' on the IUCN Red List of Threatened Species™. This medium-sized skink occurs along a narrow coastal strip of North Africa, from northeastern Morocco to northwestern Algeria, and is also found on the nearby Chafarinas Islands, Spain.

This species occupies a narrow and severely fragmented range, and is believed to be uncommon. Its habitat is likely to be under threat from coastal development, and may also be declining in quality due to the removal of driftwood for fuel, which reduces available ground cover for the species.

Doumergue's Skink occurs in a number of protected areas, including the Chafarine Hunting Reserve in the Chafarinas Islands, and the Sebkha Bou Areg and Embouchure Moulouya protected areas in Morocco. However, no specific conservation measures are currently in place for this poorly-known species.

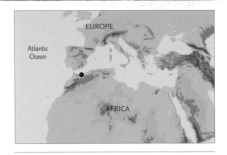

A species is **Endangered** when it faces a very high risk of extinction in the wild, based on measurements of population size and/or geographic range and their trends in the past, present and/or future.

Humphead Wrasse

Cheilinus undulatus

NOT EVALUATED	DATA DEFICIENT	LEAST CONCERN	NEAR THREATENED	VULNERABLE	⟨ ENDANGERED ⟩	CRITICALLY ENDANGERED	EXTINCT IN THE WILD	EXTINCT
NE	DD	LC	NT	VU	EN	CR	EW	EX

© CHRISTIAN LAUFENBERG

The **Humphead Wrasse**, *Cheilinus undulatus*, is listed as 'Endangered' on the IUCN Red List of Threatened Species™. It is one of the largest reef fishes in the world, reaching almost two metres in length, and occurs throughout the Indian and Pacific Oceans.

Although widespread, the Humphead Wrasse is uncommon. Its flesh is highly prized, especially in the popular live reef-fish trade. This wrasse is particularly vulnerable to exploitation; even moderate levels of fishing have a significant impact on its numbers.

Significantly, restaurants prefer the smaller, juvenile fish, so individuals are fished before they can reproduce. The Humphead Wrasse's coral reef habitat is also threatened by human activities throughout parts of its range.

The global convention to address species threatened by international trade (CITES) lists this species in its Appendix II, thus calling for regulation of trade. Fishing regulations are in place for this species in many areas, but illegal and unregulated trade persists. Tighter controls need to be implemented, particularly as this species cannot be hatchery reared, so all traded individuals come from the wild.

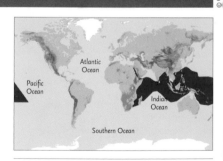

A species is **Endangered** when it faces a very high risk of extinction in the wild, based on measurements of population size and/or geographic range and their trends in the past, present and/or future.

Green Turtle

Chelonia mydas

NOT EVALUATED	DATA DEFICIENT	LEAST CONCERN	NEAR THREATENED	VULNERABLE	‹ ENDANGERED ›	CRITICALLY ENDANGERED	EXTINCT IN THE WILD	EXTINCT
NE	DD	LC	NT	VU	EN	CR	EW	EX

© XANTHE RIVETT

The **Green Turtle**, *Chelonia mydas*, is listed as 'Endangered' on the IUCN Red List of Threatened Species™. This long-lived and highly migratory species is found in tropical and, to a lesser extent, subtropical waters throughout the globe. The Green Turtle has the most numerous and widely dispersed nesting sites of the seven turtle species.

Although international trade in Green Turtles is prohibited by the Convention on International Trade in Endangered Species (CITES), Green Turtles and their eggs are still widely consumed, both legally and illegally – they were once highly sought after for their body fat, a key ingredient in the popular delicacy 'Green Turtle soup.' They are also regularly caught as by-catch in fisheries, especially by trawls, gill nets and longlines, and are threatened by coastal habitat destruction (particularly of nesting areas) and marine debris.

Green Turtles have been the focus of numerous international and regional treaties and protection measures for several decades, such as their inclusion on Appendix I of CITES. The use of turtle excluder devices in many trawl fisheries has also resulted in a decrease in incidental catch of this species, but by-catch continues to remain a significant threat globally.

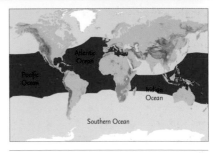

A species is Endangered when it faces a very high risk of extinction in the wild, based on measurements of population size and/or geographic range and their trends in the past, present and/or future.

Pinta Island Tortoise

Chelonoidis (nigra) abingdonii

NOT EVALUATED	DATA DEFICIENT	LEAST CONCERN	NEAR THREATENED	VULNERABLE	ENDANGERED	CRITICALLY ENDANGERED	EXTINCT IN THE WILD	EXTINCT
NE	DD	LC	NT	VU	EN	CR	EW	EX

A species is Extinct in the Wild when it is known only to survive in cultivation or in captivity. A species is presumed Extinct in the Wild when exhaustive surveys in known and/or expected habitat, at appropriate times throughout its historic range, have failed to record an individual.

© ANDERS RHODIN

The **Pinta Island Tortoise**, Chelonoidis (nigra) abingdonii, is listed as 'Extinct in the Wild' on the IUCN Red List of Threatened Species™. While there is some scientific disagreement as to whether the various Galapagos tortoises represent separate species or subspecies, all agree that Lonesome George is the last surviving individual of his kind.

The species was driven to near-extinction by collection for consumption by whalers and other Galapagos settlers, with Lonesome George being found in 1972. He was moved to the Charles Darwin Research Station in the hope that a female might be found for a captive breeding programme, but this has not happened.

Recent research, however, has demonstrated that Lonesome George's genotype is still represented among wild tortoises on Isabela Island, likely the result of a ship dropping some Pinta Island Tortoises overboard in an emergency long ago, after which some of them drifted ashore and interbred with the local tortoises. Genetic screening and selective back-crossing offers new hope that Lonesome George's lineage could be partially restored.

Burnup's Hunter Slug

Chlamydephorus burnupi

NOT EVALUATED	DATA DEFICIENT	LEAST CONCERN	NEAR THREATENED	⟨ VULNERABLE ⟩	ENDANGERED	CRITICALLY ENDANGERED	EXTINCT IN THE WILD	EXTINCT
NE	DD	LC	NT	VU	EN	CR	EW	EX

© Dai Herbert

Burnup's Hunter Slug, Chlamydephorus burnupi, is listed as 'Vulnerable' on the IUCN Red List of Threatened Species™. This species of slug is endemic to South Africa where it lives in the leaf litter of mist-belt and montane Podocarpus forests, and is largely confined to the midlands and Drakensberg foothills in KwaZulu-Natal.

This species faces various challenges in terms of its conservation, given the public view that most slugs are 'garden pests'. However, the reality is that this species has a limited range and, even within this, it is restricted to remnant pockets of suitable forest habitat. Such habitats are themselves threatened by fire, alien invasive plants, grazing and trampling by livestock, illegal harvesting of forest products and other habitat degrading processes. This slug is a predator of other invertebrates, including pill millipedes.

To date, there are no real conservation actions or interventions specifically related to Burnup's Hunter Slug. Its conservation relies very heavily on preservation of its forest habitats in KwaZulu-Natal.

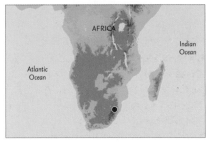

A species is Vulnerable when it faces a high risk of extinction in the wild, based on measurements of population size and/or geographic range and their trends in the past, present and/or future.

Pink Fairy Armadillo — *Chlamyphorus truncatus*

NOT EVALUATED	DATA DEFICIENT	LEAST CONCERN	NEAR THREATENED	VULNERABLE	ENDANGERED	CRITICALLY ENDANGERED	EXTINCT IN THE WILD	EXTINCT
NE	DD	LC	NT	VU	EN	CR	EW	EX

© MARIELLA SUPERINA

The **Pink Fairy Armadillo**, *Chlamyphorus truncatus*, is classified as 'Data Deficient' on the IUCN Red List of Threatened Species™. It is the smallest of all living armadillos, and can only be found in arid regions of central Argentina.

The Pink Fairy Armadillo is one of the least-known armadillos. It spends most of its time underground and is active at night, making it very difficult to observe in the wild. The reasons for the decline of its wild populations are not clear. Cattle ranching and certain agricultural practices may destroy and fragment its habitat, however, these may also cause direct harm to individuals since they dig through the sand close to the surface. Many individuals are also killed by domestic dogs and cats. In addition, this species appears to be extremely susceptible to stress.

Learning more about the biology of, and threats to, Pink Fairy Armadillos is a priority. Obtaining basic information on this species is the key to evaluating its chances of long-term survival and planning future conservation strategies.

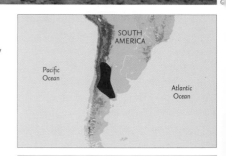

A species is **Data Deficient** when there is inadequate information to make an assessment of its risk of extinction. Listing of species in this category indicates that more information is required, and acknowledges that future research may show that threatened classification is appropriate.

Nakamura's Skydragon

Chlorogomphus nakamurai

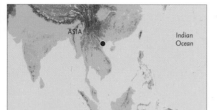

A species is **Vulnerable** when it faces a high risk of extinction in the wild, based on measurements of population size and/or geographic range and their trends in the past, present and/or future.

© Do Manh Cuong

Nakamura's Skydragon, Chlorogomphus nakamurai, is listed as 'Vulnerable' on the IUCN Red List of Threatened Species™ due to its restricted range of occurrence. This rare species was first described in 1996, and has only been found in Cuc Phuong and Ba Vi National Parks, north Vietnam.

This species has a very limited distribution in primary forests of lowland limestone areas, though the species is relatively common at the type locality. It is found at shaded, clear streams where the male usually patrols looking for females, which only come down from the tree canopies to mate and lay their eggs. Nakamura's Skydragon is threatened by habitat loss through clear-cutting of their forest habitat, particularly because timber is much easier to exploit in the lowlands than the highlands. Pollution of the forest streams will also impact on this species.

At present there are no conservation measures in place to protect Nakamura's Skydragon. However, there is a need for further sampling to determine the range of this species and to protect its forest habitat from further destruction.

Pygmy Hippopotamus

Choeropsis liberiensis

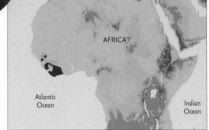

NOT EVALUATED	DATA DEFICIENT	LEAST CONCERN	NEAR THREATENED	VULNERABLE	⟨ ENDANGERED ⟩	CRITICALLY ENDANGERED	EXTINCT IN THE WILD	EXTINCT
NE	DD	LC	NT	VU	EN	CR	EW	EX

A species is **Endangered** when it faces a very high risk of extinction in the wild, based on measurements of population size and/or geographic range and their trends in the past, present and/or future.

© ZODIAC ZOOS

The **Pygmy Hippopotamus**, Choeropsis liberiensis, is listed as 'Endangered' on the IUCN Red List of Threatened Species™. Considerably less common than its larger relative, the Pygmy Hippo is restricted to the West African countries of Liberia, Côte d'Ivoire, Guinea and Sierra Leone.

Over the past 100 years, the Pygmy Hippo's habitat has declined dramatically as a result of logging, farming, and human settlement. As deforestation continues and their habitat becomes more fragmented, newly accessible populations are coming under increasing pressure from hunters. One population in Nigeria, comprising a distinct subspecies (C. liberiensis heslopi), may have already been driven to extinction.

Although a number of conservation initiatives are ongoing, without more information on the status and threats faced by this species and a coordinated conservation strategy, the Pygmy Hippo may disappear from the wild in the not-too-distant future. At present there is a good captive population, which has bred successfully and has doubled in size over the last 25 years, potentially providing a last-ditch safeguard against total extinction.

Epirus Grasshopper

NOT EVALUATED	DATA DEFICIENT	LEAST CONCERN	NEAR THREATENED	VULNERABLE	ENDANGERED	CRITICALLY ENDANGERED	EXTINCT IN THE WILD	EXTINCT
NE	DD	LC	NT	VU	EN	CR	EW	EX

© PAOLO FONTANA

The **Epirus Grasshopper**, Chorthippus lacustris, is found only in Epirus, northwestern Greece. It was discovered and described in 1975, and although it has not yet been officially classified on the IUCN Red List of Threatened Species™, its listing is potentially 'Critically Endangered'.

The species has a restricted and fragmented distribution pattern. It is known from three lake areas: the Pamvotida Lake basin, Lake Paramythia, and Lake Morfo. It is strongly dependent on wet grasslands which flood on a seasonal basis. The greatest population density is recorded in the site with the greatest diversity of dominant plant species.

The Epirus Grasshopper is estimated to have lost 85–99% of its habitat during the last 50 years due to wetland drainage. The main threat is further habitat loss as a result of urbanization around Pamvotida Lake and land conversion to agriculture around Paramythia Lake, even though both sites belong to the Natura 2000 network. Restoring wet grasslands, protecting them from further urbanization and drainage, and population monitoring are the main measures proposed for the conservation of the Epirus Grasshopper.

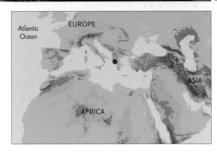

A species is Critically Endangered when it faces an extremely high risk of extinction in the wild, based on measurements of population size and/or geographic range and their trends in the past, present and/or future.

Camphor Cinnamon

Cinnamomum capparu-coronde

NOT EVALUATED	DATA DEFICIENT	LEAST CONCERN	NEAR THREATENED	‹ VULNERABLE ›	ENDANGERED	CRITICALLY ENDANGERED	EXTINCT IN THE WILD	EXTINCT
NE	DD	LC	NT	VU	EN	CR	EW	EX

The **Camphor Cinnamon**, *Cinnamomum capparu-coronde*, is listed as 'Vulnerable' on the IUCN Red List of Threatened Species™. Closely related to true Cinnamon, it is one of seven wild cinnamon species endemic to Sri Lanka. This crop wild relative grows primarily in lowland rainforests on slopes and hilly areas of altitudes between 90 m and 1,100 m.

As the local name suggests, the Camphor Cinnamon has a distinctive spicy, clove-like aroma, due to the high amounts of eugenol contained in the leaves and bark. This compound also confers medicinal properties to the plant, which is used in traditional medicine to treat a wide range of ailments including toothache, bronchitis and rheumatism.

The Camphor Cinnamon is now threatened with extinction due to habitat loss and overexploitation, despite legislation enacted in 1993 to protect wild fauna and flora such as this species. In an attempt to reverse this trend, a domestication programme has been set up by the Sri Lankan component of the Crop Wild Relative project to protect this species, which currently survives in just a few isolated populations in the wild.

© D. M. H. CHAMPIKA KUMARATHILAKE

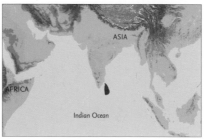

A species is Vulnerable when it faces a high risk of extinction in the wild, based on measurements of population size and/or geographic range and their trends in the past, present and/or future.

Cave Catfish

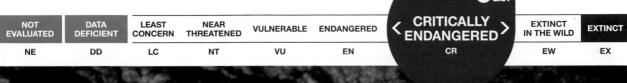

NOT EVALUATED	DATA DEFICIENT	LEAST CONCERN	NEAR THREATENED	VULNERABLE	ENDANGERED	CRITICALLY ENDANGERED	EXTINCT IN THE WILD	EXTINCT
NE	DD	LC	NT	VU	EN	CR	EW	EX

© Charles Maxwell / Underwater Video Services

The **Cave Catfish**, *Clarias cavernicola*, is listed as 'Critically Endangered' on the IUCN Red List of Threatened Species™. It is found only in the Aigamas Cave, Namibia. The pool it inhabits is in total darkness and is only 18 m by 2.5 m in area. This species has very small eyes, and is probably effectively blind.

The Cave Catfish congregate above an underwater shelf at a depth of about 15 m. The cave lake has been used as a water supply in an otherwise very dry area, and the pumping of water has reduced the depth of the lake from 70 m to 50 m. The shelf where the catfish are found is in danger of being exposed due to the decreasing water levels. The potential threat of illegal collecting for the aquarium trade also exists.

This is thought to be one of the most threatened fishes in the world, and the control of water extraction to protect its habitat is desperately needed. Attempts to breed the Cave Catfish in captivity have failed, and any further collection of this species should be tightly regulated by permits.

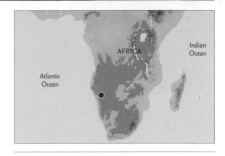

A species is Critically Endangered when it faces an extremely high risk of extinction in the wild, based on measurements of population size and/or geographic range and their trends in the past, present and/or future.

NOT EVALUATED	DATA DEFICIENT	LEAST CONCERN	>	NEAR THREATENED	VULNERABLE	ENDANGERED	CRITICALLY ENDANGERED	EXTINCT IN THE WILD	EXTINCT
NE	DD	LC		NT	VU	EN	CR	EW	EX

Molina's Hog-nosed Skunk, *Conepatus chinga*, is listed as 'Least Concern' on the IUCN Red List of Threatened Species™. It inhabits open savannahs and arid and shrubby areas from southern Peru, through Bolivia, and south to Uruguay, western Paraguay, central Chile and Argentina. It also occurs in several locations in southern Brazil.

During the 1970s and early 1980s, skunks in Argentina were heavily hunted for their fur. Large areas of suitable habitat have also been severely degraded, due to feral introduced species and overgrazing by livestock which lead to increased soil erosion. However, Molina's Hog-nosed Skunk is thought to remain widespread and locally common, and is not believed to be undergoing a rapid decline.

A range of conservation actions have been recommended for native species in the Patagonian Steppe, including the creation of protected areas where livestock are excluded, and the prevention of new exotic species introductions. It has also been recommended that *Conepatus* species be included on Appendix II of CITES, in order to better estimate and regulate the levels of exploitation of these species.

© CARLOS BENHUR KASPER

A species is **Least Concern** when it does not qualify for Critically Endangered, Endangered, Vulnerable or Near Threatened. Widespread and abundant species are included in this category.

NOT EVALUATED	DATA DEFICIENT	LEAST CONCERN	NEAR THREATENED	VULNERABLE	⟨ ENDANGERED ⟩	CRITICALLY ENDANGERED	EXTINCT IN THE WILD	EXTINCT
NE	DD	LC	NT	VU	EN	CR	EW	EX

© Ignacio De La Riva

The **Goliath Frog**, *Conraua goliath*, is listed as 'Endangered' on the IUCN Red List of Threatened Species™. This aptly named species is the world's largest frog and is restricted to a narrow region in west-central Africa.

With trapping methods becoming increasingly sophisticated, harvesting for food has caused the Goliath Frog population to decline. The enormous size of this species has also made it a target for the pet trade, with approximately 300 individuals imported from Cameroon to the United States each year. Additionally, the Goliath Frog is affected by the loss of forest habitat for agriculture, logging, and human settlements, as well as by sedimentation of its breeding streams.

To conserve the Goliath Frog, efforts need to be made to safeguard areas of remaining habitat, and to work with local communities to manage harvests at sustainable levels. If the population continues to decline, a captive breeding programme could be critical to the survival of this species in the long-term.

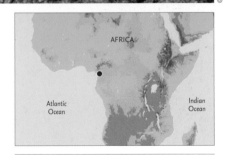

A species is **Endangered** when it faces a very high risk of extinction in the wild, based on measurements of population size and/or geographic range and their trends in the past, present and/or future.

NOT EVALUATED	DATA DEFICIENT	LEAST CONCERN	NEAR THREATENED	VULNERABLE	ENDANGERED	CRITICALLY ENDANGERED	EXTINCT IN THE WILD	EXTINCT
NE	DD	LC	NT	VU	EN	CR	EW	EX

© NICK HELME

Cotula myriophylloides is listed as 'Critically Endangered' on the IUCN Red List of Threatened Species™. This bright yellow plant from the daisy family is endemic to the Cape Peninsula in the Western Cape, South Africa, and may only be present in one location (Noordhoek). It grows in still or slowly moving fresh- and brackish water, sometimes forming floating mats in marshes and on wet sand.

The main threats to this species are urbanization and invasive alien species. Urbanization and development (such as the Noordhoek Saltpans, which have been developed over the past 10 years)

have caused drainage of wetlands, and many historically seasonal wetlands are now permanent eutrophic wetlands. These wetlands are often invaded by alien species that compete with native species.

Cotula myriophylloides has not been seen for the past 30 years, even though its range falls within the Cape Peninsula, which is a well-sampled area. This species is very likely to be extinct, thus urgent surveys are needed to confirm if it is still in existence.

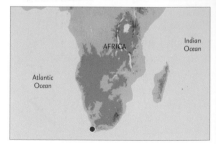

A species is **Critically Endangered** when it faces an extremely high risk of extinction in the wild, based on measurements of population size and/or geographic range and their trends in the past, present and/or future.

Kitti's Hog-nosed Bat

Craseonycteris thonglongyai

NOT EVALUATED	DATA DEFICIENT	LEAST CONCERN	NEAR THREATENED	⟨ VULNERABLE ⟩	ENDANGERED	CRITICALLY ENDANGERED	EXTINCT IN THE WILD	EXTINCT
NE	DD	LC	NT	VU	EN	CR	EW	EX

Kitti's Hog-nosed Bat, *Craseonycteris thonglongyai*, is listed as 'Vulnerable' on the IUCN Red List of Threatened Species™. Also known as the Bumblebee Bat, this tiny species from west-central Thailand and southeast Myanmar, is the world's smallest known mammal and the only member of the family Craseonycteridae.

Since its discovery in the 1970s, Kitti's Hog-nosed Bat has come under threat from human disturbance, fertilizer collection and tourism. Another serious threat arises from the burning of forest near bat caves, which destroys critical foraging habitat, while proposals to extract limestone from caves inhabited by this species could also be highly destructive.

Kitti's Hog-nosed Bats occur within Sai Yok National Park, which offers some protection. A conservation action plan was created for the species in 2001, and recommends actions for its conservation. These recommendations include monitoring, providing incentives to local people to maintain essential habitat, and identifying and protecting key cave roosts.

© MERLIN D. TUTTLE, BAT CONSERVATION INTERNATIONAL, WWW.BATCON.ORG

A species is **Vulnerable** when it faces a **high risk** of extinction in the wild, based on measurements of population size and/or geographic range and their trends in the past, present and/or future.

Miles' Robber Frog

Craugastor milesi

NOT EVALUATED	DATA DEFICIENT	LEAST CONCERN	NEAR THREATENED	VULNERABLE	ENDANGERED	CRITICALLY ENDANGERED	EXTINCT IN THE WILD	EXTINCT
NE	DD	LC	NT	VU	EN	CR	EW	EX

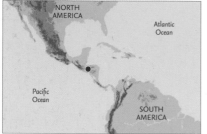

A species is **Critically Endangered** when it faces an extremely high risk of extinction in the wild, based on measurements of population size and/or geographic range and their trends in the past, present and/or future.

© JAMES McCRANIE

Miles' Robber Frog, *Craugastor milesi*, is listed as 'Critically Endangered' on the IUCN Red List of Threatened Species™. This Honduran endemic was once considered 'Extinct in the Wild' as it had not been seen since 1983, and surveys between 1992 and 1998 failed to find it.

However, the species was rediscovered in 2008 after a single individual was found at Cusuco National Park, western Honduras.

Once abundant, Miles' Robber Frog declined dramatically throughout the 1980s. Habitat loss and conversion likely impacted upon the species, but the sudden disappearance of populations in areas of pristine forest may be attributed to chytridiomycosis. This fungus is known to infect other amphibian species in the region, and the species' affinity to streamside habitats makes it highly susceptible to infection.

The recent rediscovery of Miles' Robber Frog gives rise to the hope that a single resistant population remains. However, the species' status is still extremely precarious, and research is urgently required to determine the viability of the surviving population.

Philippine Crocodile

NOT EVALUATED	DATA DEFICIENT	LEAST CONCERN	NEAR THREATENED	VULNERABLE	ENDANGERED	CRITICALLY ENDANGERED	EXTINCT IN THE WILD	EXTINCT
NE	DD	LC	NT	VU	EN	CR	EW	EX

© MERLIJN VAN WEERD

The **Philippine Crocodile**, *Crocodylus mindorensis*, is listed as 'Critically Endangered' on the IUCN Red List of Threatened Species™. Found only in the Philippines, this relatively small crocodile is now reduced to just two viable populations on Luzon and Mindanao Islands, with a few small groups elsewhere.

The species' initial decline was due to commercial overexploitation, followed by widespread habitat destruction. The decline continued as wetlands were converted for other uses, and the hunting of crocodiles persisted.

Recommended by the Crocodile Specialist Group, a successful captive breeding programme was initiated on Palawan Island in 1987. The Philippine Government is implementing a national recovery plan, which was produced in 2000 with extensive community input. Community-based conservation programmes are in place for the Mindanao and Luzon populations. Habitat protection, coupled with regular monitoring and increasing community support, has halted the species' decline, and the Philippine Crocodile's future is looking more positive. However, with fragile social environments and an estimated total wild population of perhaps 350 individuals, much work remains in order to enable this species to recover fully.

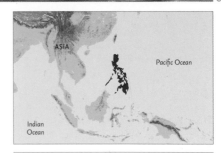

A species is *Critically Endangered* when it faces an extremely high risk of extinction in the wild, based on measurements of population size and/or geographic range and their trends in the past, present and/or future.

Santa Catalina Island Rattlesnake

NOT EVALUATED	DATA DEFICIENT	LEAST CONCERN	NEAR THREATENED	VULNERABLE	ENDANGERED	CRITICALLY ENDANGERED	EXTINCT IN THE WILD	EXTINCT
NE	DD	LC	NT	VU	EN	CR	EW	EX

The **Santa Catalina Island Rattlesnake**, *Crotalus catalinensis*, is listed as 'Critically Endangered' on the IUCN Red List of Threatened Species™. As its name suggests, this rattlesnake is restricted to Isla Santa Catalina in the Gulf of California, an island covering just 40 square kilometres.

Once thought to be a common species, the Santa Catalina Island Rattlesnake has suffered declines, primarily due to the killing and illegal collection of individuals. Unfortunately, its passive behaviour makes it an easy target. In addition, a decline in the Santa Catalina Island Rattlesnake's main prey, the Deer Mouse, may pose a further risk to this highly threatened species.

Like many snakes, the Santa Catalina Island Rattlesnake may receive less conservation attention than it deserves due to long-standing and fairly widespread negative attitudes towards snakes. There is a need to monitor populations and to prevent over-collection of this restricted range species.

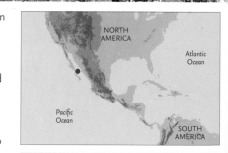

A species is **Critically Endangered** when it faces an extremely high risk of extinction in the wild, based on measurements of population size and/or geographic range and their trends in the past, present and/or future.

Root-spine Palm

Cryosophila williamsii

NOT EVALUATED	DATA DEFICIENT	LEAST CONCERN	NEAR THREATENED	VULNERABLE	ENDANGERED	CRITICALLY ENDANGERED	EXTINCT IN THE WILD	EXTINCT
NE	DD	LC	NT	VU	EN	CR	EW	EX

© Scott Zona

A species is **Extinct in the Wild** when it is known only to survive in cultivation or in captivity. A species is presumed Extinct in the Wild when exhaustive surveys in known and/or expected habitat, at appropriate times throughout its historic range, have failed to record an individual.

The **Root-spine Palm**, Cryosophila williamsii, is listed as 'Extinct in the Wild' on the IUCN Red List of Threatened Species™. Restricted to a small area of west-central Honduras, this solitary palm was formerly found in rainforest on steep, high-rainfall slopes of the Lago Yojoa watershed.

The greatest threat to the Root-spine Palm is deforestation due to increasing agriculture, settlement and logging. The seedlings require moist shade to establish and grow, and are unable to tolerate the exposure that occurs after heavy deforestation. In addition, this species has been exploited for thatch and for its palm hearts.

The Lago Yojoa watershed is a designated forest reserve, but deforestation continues there. Surveys are needed to establish whether the Root-spine Palm is truly extinct in the wild and, if any individuals remain, urgent protection will be needed.

The greatest hope for the species is now an ex situ collection at Fairchild Tropical Botanical Garden; if seeds can be produced from these cultivated plants, reintroduction efforts should be attempted, and additional ex situ collections should also be established.

NOT EVALUATED	DATA DEFICIENT	LEAST CONCERN	NEAR THREATENED	VULNERABLE	ENDANGERED	‹ CRITICALLY ENDANGERED ›	EXTINCT IN THE WILD	EXTINCT
NE	DD	LC	NT	VU	EN	CR	EW	EX

© David Harries

Cryptomyces maximus is currently being assessed for the IUCN Red List of Threatened Species™, however it has a provisional assessment of 'Critically Endangered'. This fungus has been known for over 200 years, but has never been common and is now, in fact, exceptionally rare. Searches throughout Europe since 2000 for this conspicuous fungus revealed sites only in southwest Wales and northern Sweden.

Cryptomyces maximus occurs only on willows, as a weak parasite of twigs, probably only those already damaged (for example by ponies). In this highly specialized habitat, Cryptomyces maximus does not seriously threaten the tree. It produces extensive irregular black crusts below the bark, characteristically surrounded by a bright orange or yellow halo. When ripe and moist, the fruitbody expands, breaking through the bark, and splitting the crust to reveal the fertile layer. Ascospores (sexually produced microscopic fungal spores) are released violently from this surface and are dispersed by the wind.

Although part of the Pembroke Coast National Park, the area in Wales with known colonies is very small, making Cryptomyces maximus exceptionally prone to single catastrophic events.

A species is **Critically Endangered** when it faces an extremely high risk of extinction in the wild, based on measurements of population size and/or geographic range and their trends in the past, present and/or future.

Fossa

Cryptoprocta ferox

NOT EVALUATED	DATA DEFICIENT	LEAST CONCERN	NEAR THREATENED	‹ VULNERABLE ›	ENDANGERED	CRITICALLY ENDANGERED	EXTINCT IN THE WILD	EXTINCT
NE	DD	LC	NT	VU	EN	CR	EW	EX

The **Fossa**, *Cryptoprocta ferox*, is listed as 'Vulnerable' on the IUCN Red List of Threatened Species™. It is the largest mammalian carnivore on Madagascar, and has a wide distribution in the less disturbed forests across the island.

The main threat to the Fossa is the loss and fragmentation of its habitat, as forests are selectively logged or converted for agriculture and pasture. The species is sometimes killed by local people as a pest of domestic poultry, and is sometimes taken for its body parts, which are used for medicinal purposes. Competition with introduced carnivores such as feral dogs may also be a threat.

The Fossa is present in many protected areas throughout Madagascar, and also receives some protection from international trade under its listing on Appendix II of CITES. A successful captive breeding programme is in place for the species, but better protection of its forest habitat is needed, together with awareness programmes to highlight its value in controlling pests, and improved protection under national legislation.

© LUKE DOLLAR

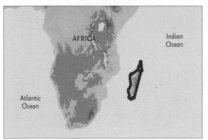

A species is Vulnerable when it faces a high risk of extinction in the wild, based on measurements of population size and/or geographic range and their trends in the past, present and/or future.

Utila Spiny-tailed Iguana

Ctenosaura bakeri

NOT EVALUATED	DATA DEFICIENT	LEAST CONCERN	NEAR THREATENED	VULNERABLE	ENDANGERED	CRITICALLY ENDANGERED	EXTINCT IN THE WILD	EXTINCT
NE	DD	LC	NT	VU	EN	CR	EW	EX

The **Utila Spiny-tailed Iguana**, or Swamper, *Ctenosaura bakeri*, is classified as 'Critically Endangered' on the IUCN Red List of Threatened Species™. Once considered to be one of the rarest iguanas in existence, the Utila Spiny-tailed Iguana is named after the Caribbean island of Utila, Honduras, which it inhabits, and the whorls of enlarged spiny scales that encircle the tail.

Declared functionally extinct in the early nineties, there is now thought to be a population of approximately 10,000 in two or three groups. Distribution is linked to habitat where sandy, open nest sites, as well as mangroves, are essential. The major threats to the Utila Spiny-tailed Iguana are tourist development, deforestation for local timber use, invasive plants which cover nest sites, hunting and rubbish dumping.

In 1994, the Frankfurt Zoological Society and the Senckenberg Nature Research Society began protecting the species. The Iguana Research and Breeding Station has since developed, which has been very successful in releasing iguanas into suitable remaining habitat. Despite these efforts, further work is still needed to help to protect the Utila Spiny-tailed Iguana and its fragile mangrove habitat.

A species is Critically Endangered when it faces an extremely high risk of extinction in the wild, based on measurements of population size and/or geographic range and their trends in the past, present and/or future.

Dhole

Cuon alpinus

NOT EVALUATED	DATA DEFICIENT	LEAST CONCERN	NEAR THREATENED	VULNERABLE	‹ ENDANGERED ›	CRITICALLY ENDANGERED	EXTINCT IN THE WILD	EXTINCT
NE	DD	LC	NT	VU	EN	CR	EW	EX

© Bruce Kekule / BRUCEKEKULE.COM

The **Dhole**, *Cuon alpinus*, is listed as 'Endangered' on the IUCN Red List of Threatened Species™. This large, social canid, also called the Asiatic Wild Dog, once ranged throughout the Indian subcontinent, north into Korea, China and eastern Russia, and south through Malaysia and Indonesia, reaching as far as Java. However, its current range is greatly reduced and highly fragmented.

There are estimated to be fewer than 2,500 Dholes left in the wild. While hunting the species is legally prohibited, Dholes are viewed as a menace to humans and their livestock, so are persecuted by trapping, shooting, and poisoning. The most prominent threats facing the species are the widespread loss of habitat and the depletion of its main prey base (deer) due to excessive hunting.

Dholes have been reported to occur in a number of protected areas throughout their disjunct distribution. The Dhole Conservation Project is working to gather more data on the species' distribution and population status. Additional work is needed to understand the potential threat posed by domestic or feral dogs as vectors of pathogens and disease.

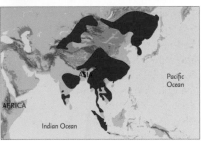

A species is **Endangered** when it faces a very high risk of extinction in the wild, based on measurements of population size and/or geographic range and their trends in the past, present and/or future.

Yunnan Box Turtle

Cuora yunnanensis

NOT EVALUATED	DATA DEFICIENT	LEAST CONCERN	NEAR THREATENED	VULNERABLE	ENDANGERED	CRITICALLY ENDANGERED	EXTINCT IN THE WILD	EXTINCT
NE	DD	LC	NT	VU	EN	CR	EW	EX

The **Yunnan Box Turtle**, *Cuora yunnanensis*, is listed as 'Critically Endangered' on the IUCN Red List of Threatened Species™, a decade after being listed as 'Extinct'. This species was first discovered around 1906 in high altitude areas in China, and was not recorded for many years until a few captive animals appeared in 2005.

There were doubts as to whether these captive-bred animals were genuine Yunnan Box Turtles, but genetic testing showed that they were. The reasons for the species' rarity are not clear, but are thought to include the limited availability of suitable habitats at high elevations and collection for consumption and medicinal use, as well as widespread habitat destruction and degradation.

While potentially extremely valuable in the global pet trade – the first animal to emerge from China into the international pet trade is reputed to be worth about USD 50,000 – the Yunnan Box Turtle is strictly protected in China, where some of the known animals are safeguarded in captivity. A comprehensive captive breeding and species recovery programme is also taking shape.

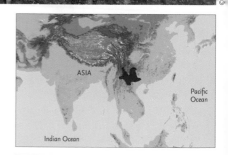

A species is **Critically Endangered** when it faces an extremely high risk of extinction in the wild, based on measurements of population size and/or geographic range and their trends in the past, present and/or future.

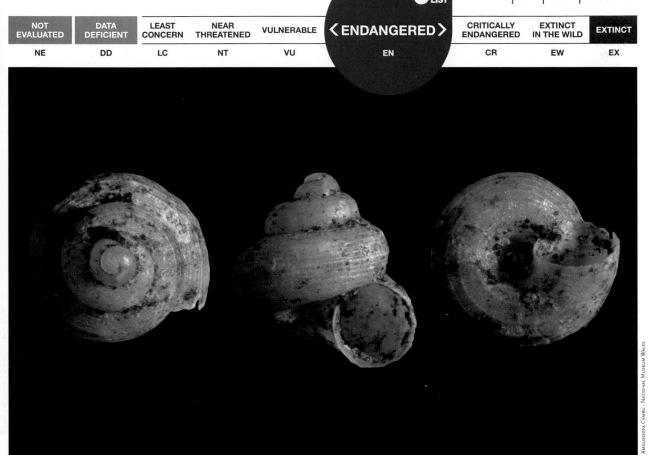

NOT EVALUATED	DATA DEFICIENT	LEAST CONCERN	NEAR THREATENED	VULNERABLE	⟨ ENDANGERED ⟩	CRITICALLY ENDANGERED	EXTINCT IN THE WILD	EXTINCT
NE	DD	LC	NT	VU	EN	CR	EW	EX

© Amgueddfa Cymru – National Museum Wales

Cyathopoma picardense is listed as 'Endangered' on the IUCN Red List of Threatened Species™. Endemic to Aldabra Atoll in the Seychelles, this land snail inhabits coastal scrub and woodland from sea level up to six metres above sea level.

As most of Aldabra Atoll is only 1–2 metres above sea level, Cyathopoma picardense is under huge threat from sea level rise. Although no population estimates are currently available, the population is generally believed to be decreasing.

Aldabra Atoll is well protected, being designated a Special Reserve in 1976 and a World Heritage Site in 1982. Nonetheless, research is needed to study the life history, ecology and population status of Cyathopoma picardense, and its habitat needs to be monitored. Given the possible effects of climate change, artificial propagation of this species should be considered, and the species' genes should be preserved in a genome resource bank.

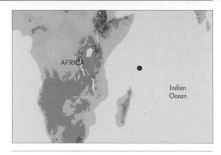

A species is **Endangered** when it faces a very high risk of extinction in the wild, based on measurements of population size and/or geographic range and their trends in the past, present and/or future.

NOT EVALUATED	DATA DEFICIENT	LEAST CONCERN	NEAR THREATENED	VULNERABLE	‹ ENDANGERED ›	CRITICALLY ENDANGERED	EXTINCT IN THE WILD	EXTINCT
NE	DD	LC	NT	VU	EN	CR	EW	EX

RED LIST

© THOMAS MARLER

Cycas micronesica is listed as 'Endangered' on the IUCN Red List of Threatened Species™. The plant occurs in remnant native forest on the islands of Guam, Rota, Yap and Palau in the western Pacific Ocean.

The major historical threat was habitat destruction, but despite this, the species was the most abundant tree on Guam at the end of the 20th century. However, the scale Aulacaspis yasumatsui invaded Guam in 2003 and the butterfly Chilades pandava followed in 2005. These cycad-specific pests have combined with other herbivores to eliminate seedling and juvenile trees and reduce the population of mature trees by 75%. There is no recruitment due to reduced seed production and low germination, combined with scale infestations that kill all new seedlings. The scale pest has spread throughout all the cycad-inhabited forests of Guam, and onto the islands of Rota and Palau.

Several management approaches have had limited success, including biological and chemical control of the scale. In case this species becomes locally extinct on Guam and Rota, three ex situ cycad gardens have been established with seeds from several different habitats around Guam.

A species is **Endangered** when it faces a very high risk of extinction in the wild, based on measurements of population size and/or geographic range and their trends in the past, present and/or future.

Silky Anteater

Cyclopes didactylus

NOT EVALUATED	DATA DEFICIENT	LEAST CONCERN	NEAR THREATENED	VULNERABLE	ENDANGERED	CRITICALLY ENDANGERED	EXTINCT IN THE WILD	EXTINCT
NE	DD	LC	NT	VU	EN	CR	EW	EX

The **Silky Anteater**, Cyclopes didactylus, is classified as 'Least Concern' on the IUCN Red List of Threatened Species™. This is the smallest of all anteaters, weighing only 300 grammes as an adult. It is extremely difficult to observe in the wild because it is only active at night and moves through the canopy without descending to the ground.

The Silky Anteater is distributed throughout the Amazonian rainforest. An isolated, genetically distinct population exists in the northeastern Atlantic Forest of Brazil and is regionally classified as 'Critically Endangered'. This population is particularly threatened by habitat destruction, as deforestation is advancing at a fast pace. Locals also frequently capture Silky Anteaters to keep them as pets.

Field research has been initiated by the Brazilian NGO Projeto Tamandua to obtain basic information on this poorly known species and to promote its conservation. Education programmes are urgently needed to stop its use as a pet, and effective conservation of its natural forest habitat is the key to ensuring the long-term survival of this species.

© FLÁVIA MIRANDA

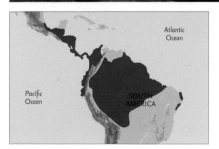

A species is **Least Concern** when it does not qualify for Critically Endangered, Endangered, Vulnerable or Near Threatened. Widespread and abundant species are included in this category.

Jamaican Iguana

Cyclura collei

The **Jamaican Iguana**, *Cyclura collei*, is listed as 'Critically Endangered' on the IUCN Red List of Threatened Species™. It was considered extinct for much of the last century, until its rediscovery in 1990. Formerly distributed along much of the island's dry south coast, this species is among the rarest of the world's lizards, and persists only in the rugged Hellshire Hills.

Much of its historical habitat was lost to agriculture and urban development, but the introduction of non-native mammalian predators, particularly the mongoose, is considered to be the main threat to the species. There are thought to be as few as 100 adult Jamaican Iguanas left.

The species is now supported by a continuous recovery programme that includes rearing young and releasing them as juveniles, and predator control near main nesting sites. To date, over 100 individuals have been repatriated into Hellshire, and those animals are increasing the nesting population in the core conservation zone. In fact, the number of females using the two known communal nesting areas has more than doubled in the past 20 years.

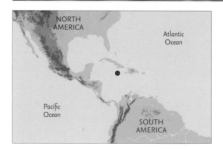

A species is **Critically Endangered** when it faces an extremely high risk of extinction in the wild, based on measurements of population size and/or geographic range and their trends in the past, present and/or future.

Anegada Ground Iguana

Cyclura pinguis

NOT EVALUATED	DATA DEFICIENT	LEAST CONCERN	NEAR THREATENED	VULNERABLE	ENDANGERED	CRITICALLY ENDANGERED	EXTINCT IN THE WILD	EXTINCT
NE	DD	LC	NT	VU	EN	CR	EW	EX

© Glenn Gerber

The **Anegada Ground Iguana**, *Cyclura pinguis*, is listed as 'Critically Endangered' on the IUCN Red List of Threatened Species™. Once distributed across the entire Puerto Rico Bank, this species is now restricted to the island of Anegada, with introduced populations on the islands of Guana, Necker, Norman and Little Thatch.

Human encroachment and the introduction of non-native mammals, especially feral cats, have caused the decline of this species. Each autumn as hatchling iguanas emerge from their nests, feral cats prey on the naïve iguanas resulting in high juvenile mortality. Today the wild population is made up almost entirely of older adults; perhaps fewer than 400 Anegada Ground Iguanas remain in the wild.

A range of conservation activities are underway on Anegada, including establishing a proposed national park, encompassing the core iguana area, and a 'headstart' facility which raises hatchling iguanas in captivity until they are large enough to be safely released. To date, over 100 iguanas have been released back into the wild with an 80% survival rate.

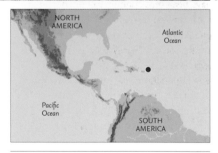

A species is **Critically Endangered** when it faces an extremely high risk of extinction in the wild, based on measurements of population size and/or geographic range and their trends in the past, present and/or future.

Otter Civet

Cynogale bennettii

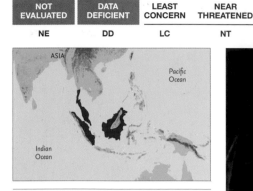

A species is **Endangered** when it faces a very high risk of extinction in the wild, based on measurements of population size and/or geographic range and their trends in the past, present and/or future.

The **Otter Civet**, Cynogale bennettii, is listed as 'Endangered' on the IUCN Red List of Threatened Species™. As its name suggests, the Otter Civet is a semi-aquatic species with several adaptations to life in water, including dense hair, valve-like nostrils and webbed feet.

It is known from Malaysia, Indonesia and Thailand, and allegedly also northern Vietnam and southern Yunnan, China.

Owing to habitat destruction and pollution, the Otter Civet population is estimated to have undergone a 50% decline over the last 15 years. Clear-cut logging is one of the major factors contributing to the decline in suitable habitat, while the conversion of peat swamp forests to oil palm plantations is another major threat.

Conservation of this rare species primarily requires protection of forest and riverine habitat, and policing against illegal harvesting of timber. In addition, further research needs to be carried out to assess its distribution and monitor its population over time. Fortunately, the Otter Civet does already occur in several protected areas throughout its range.

Hooded Seal

Cystophora cristata

A species is Vulnerable when it faces a high risk of extinction in the wild, based on measurements of population size and/or geographic range and their trends in the past, present and/or future.

© Kit M. Kovacs & Christian Lydersen

The **Hooded Seal**, *Cystophora cristata*, is listed as 'Vulnerable' on the IUCN Red List of Threatened Species™. This species gets its name from the striking nasal appendage found on adult males. Hooded Seals live at high latitudes in the North Atlantic, where they breed on drifting pack ice before dispersing in northern waters.

Hooded Seals have been commercially hunted for centuries and at times hunting levels have been unsustainable. Currently, the Hooded Seal population in the Northwest Atlantic appears to be stable, but the Northeast Atlantic population has declined markedly. Commercial hunting has likely contributed to the 85% reduction in this population in recent decades, and deterioration of ice conditions within the breeding area is almost certainly contributing. The collapse of stocks of favoured prey such as redfish may also be a factor.

Harvesting nations have employed various conservation measures to achieve sustainable hunting and allow overharvested populations to recover. These include quota regulations and limitations on hunting gear and seasons. Additionally, international management plans, agreements and treaties have been implemented to conserve populations.

Mahoenui Giant Weta

Deinacrida mahoenui

NOT EVALUATED	DATA DEFICIENT	LEAST CONCERN	NEAR THREATENED	VULNERABLE	⟨ ENDANGERED ⟩	CRITICALLY ENDANGERED	EXTINCT IN THE WILD	EXTINCT
NE	DD	LC	NT	VU	EN	CR	EW	EX

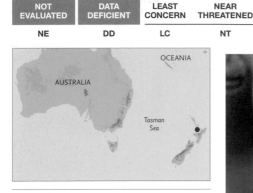

A species is **Endangered** when it faces a *very high* risk of extinction in the wild, based on measurements of population size and/or geographic range and their trends in the past, present and/or future.

The **Mahoenui Giant Weta**, *Deinacrida mahoenui*, is not yet officially listed on the IUCN Red List of Threatened Species™, however its conservation status is potentially 'Endangered'. It was not discovered until the 1960s and is found only in New Zealand.

This flightless arboreal insect was once widespread throughout the forests of the central North Island of New Zealand. However, only one population survived in a patch of gorse, considered a weed. This prickly plant provided protection for the Mahoenui Giant Weta from introduced mammals, such as rats, hedgehogs and possums.

To reduce the risk of extinction, translocations of this gentle giant were made because this species is known only at a single location where there is a high fire risk. Research has found the translocated weta flourishing at only two locations, and these new populations have successfully become established in the absence of introduced mammals, particularly rats. Future efforts to establish Mahoenui Giant Weta should be in mammal-free sanctuaries, and long-term monitoring of all populations is required, particularly at sites where the likelihood of mammal reinvasion is high.

NOT EVALUATED	DATA DEFICIENT	LEAST CONCERN	NEAR THREATENED	⟨ VULNERABLE ⟩	ENDANGERED	CRITICALLY ENDANGERED	EXTINCT IN THE WILD	EXTINCT
NE	DD	LC	NT	VU	EN	CR	EW	EX

© Danny Thornburrow, Landcare Research, NZ

The **Stephens Island Weta**, *Deinacrida rugosa*, is classified as 'Vulnerable' on the IUCN Red List of Threatened Species™. The Stephens Island Weta is endemic to New Zealand. They are nocturnal, large-bodied and flightless insects, and live for two to three years.

Once found throughout the lower North Island of New Zealand, this harmless giant has succumbed to the invasion of introduced mammals such as rats, cats and hedgehogs. They now only survive on a few predator-free islands, away from the North Island mainland.

Since 1977, the Stephens Island Weta has been translocated to four new mammal-free islands, both for conservation of the species and as part of island restoration programmes. Recently, some weta were translocated into a fenced mainland area where all mammalian pests (except mice) had been successfully removed, restoring it to the main North Island, where they had been extinct for over 100 years. The Stephens Island Weta remain highly vulnerable to the reinvasion of introduced mammals, and ongoing monitoring and management will be required in order to safeguard the future of these gentle giants.

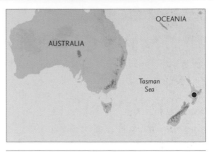

A species is Vulnerable when it faces a high risk of extinction in the wild, based on measurements of population size and/or geographic range and their trends in the past, present and/or future.

Central American River Turtle

NOT EVALUATED	DATA DEFICIENT	LEAST CONCERN	NEAR THREATENED	VULNERABLE	ENDANGERED	CRITICALLY ENDANGERED	EXTINCT IN THE WILD	EXTINCT
NE	DD	LC	NT	VU	EN	CR	EW	EX

© JOHN POLISAR

The **Central American River Turtle**, *Dermatemys mawii*, is listed as 'Critically Endangered' on the IUCN Red List of Threatened Species™. The last remaining representative of a turtle family dating back 65 million years, this unique species reaches a shell length of up to 60 cm.

It is entirely aquatic, inhabiting rivers, lagoons, and other large wetlands in southern Mexico, Guatemala, and Belize. It is barely able to move on land, and rather than emerging onto land to nest, females instead construct nests at the waterline during floods, and the eggs only begin developing after the water level has dropped. This species is highly esteemed for local consumption, and intensive collection, particularly ahead of festivals, has depleted populations severely across most or all of its range.

It is protected or regulated in all range countries and under CITES, but local enforcement is largely lacking. Increased public awareness, population recovery efforts, and improved enforcement to prevent overexploitation are all urgently needed.

A species is **Critically Endangered** when it faces an extremely high risk of extinction in the wild, based on measurements of population size and/or geographic range and their trends in the past, present and/or future.

Leatherback Turtle

NOT EVALUATED	DATA DEFICIENT	LEAST CONCERN	NEAR THREATENED	VULNERABLE	ENDANGERED	CRITICALLY ENDANGERED	EXTINCT IN THE WILD	EXTINCT
NE	DD	LC	NT	VU	EN	CR	EW	EX

© BRIAN J HUTCHINSON

The **Leatherback Turtle**, Dermochelys coriacea, is listed as 'Critically Endangered' on the IUCN Red List of Threatened Species™. It is the world's largest turtle and is found throughout every ocean (it has been recorded as far north as Alaska and as far south as the tip of South Africa).

Threats to Leatherback Turtles worldwide include loss of nesting habitat, accidental capture in fishing lines and nets, collisions with boats, egg collection for human consumption, and ingestion of discarded plastics which are often mistaken for jellyfish (their preferred diet).

Exploitation of sea turtles and their products has become illegal in most countries. Conservation programmes have been established in most of their nesting areas to protect egg clutches and nesting females from poachers. Because of the severe decline in the world's population of Leatherback Turtles, better protection of critical nesting habitat, and the reduction of incidental captures in fisheries, is essential. Furthermore, because the migratory routes of this species cross territorial waters of many nations, further international collaboration focusing on conservation will greatly enhance its chances of survival.

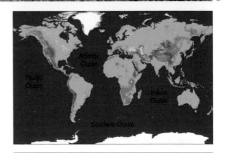

A species is Critically Endangered when it faces an extremely high risk of extinction in the wild, based on measurements of population size and/or geographic range and their trends in the past, present and/or future.

NOT EVALUATED	DATA DEFICIENT	LEAST CONCERN	NEAR THREATENED	‹ VULNERABLE ›	ENDANGERED	CRITICALLY ENDANGERED	EXTINCT IN THE WILD	EXTINCT
NE	DD	LC	NT	VU	EN	CR	EW	EX

The **Russian Desman**, *Desmana moschata*, is listed as 'Vulnerable' on the IUCN Red List of Threatened Species™. This small insectivore belongs to the same family as moles, but is adapted to a more aquatic lifestyle. A rare species, having recently gone extinct in Belarus, it now only occurs in Russia, Ukraine and Kazakhstan.

During the late 19th and early 20th centuries, the Russian Desman was massively exploited for its fur and musk glands, and as a result the population was decimated. Although this species is now fully protected and the fur trade no longer poses a threat, large numbers of desmans are accidently trapped in fishing nets. Additional threats include overfishing of prey stock, habitat loss and degradation, water pollution, and competition from introduced species.

A number of reserves have been established to protect the last unspoilt wetlands within the Russian Desman's range. Given the threat posed by current fishing practices, campaigns are underway to ban the use of nylon nets and electric landing nets.

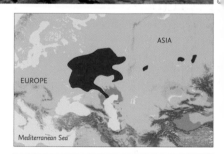

A species is Vulnerable when it faces a high risk of extinction in the wild, based on measurements of population size and/or geographic range and their trends in the past, present and/or future.

NOT EVALUATED	DATA DEFICIENT	LEAST CONCERN	NEAR THREATENED	VULNERABLE	ENDANGERED	CRITICALLY ENDANGERED	EXTINCT IN THE WILD	EXTINCT
NE	DD	LC	NT	VU	EN	CR	EW	EX

© ALAIN MICHAUD

Diacheopsis metallica has not yet been assessed for the IUCN Red List of Threatened Species™, however it has a provisional listing of 'Near Threatened'. This species is a myxomycete (also known as slime moulds), which are protozoans that look like fungi for one part of their life cycle, but like amoebae at other times. Myxomycetes inhabit rotten wood, bark, leaf litter and other decaying organic matter. They are thought to feed on dead animal, fungal and plant remains, and living and dead bacteria.

This species produces small cushion-shaped or subglobose (not quite the perfect shape of a sphere) fruitbodies with a spectacular shiny metallic sheen on the surface. These fruitbodies split open, liberating spiny spores which are dispersed by insects, water and wind. Diacheopsis metallica belongs in a special ecological group called the nivicolous myxomycetes. These are found only in the spring on mountains, typically with high insolation near melting snow.

The snow cover prevents abrupt soil temperature changes, and provides free water and a favourable ground-level microclimate. Nivicolous myxomycetes are threatened by diminishing snow cover resulting from climate change. There are currently no conservation actions for Diacheopsis metallica.

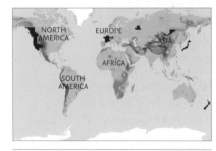

A species is Near Threatened when it does not qualify for Critically Endangered, Endangered or Vulnerable now, but is close to qualifying for or is likely to qualify for a threatened category in the near future.

Black Rhinoceros

NOT EVALUATED	DATA DEFICIENT	LEAST CONCERN	NEAR THREATENED	VULNERABLE	ENDANGERED	CRITICALLY ENDANGERED	EXTINCT IN THE WILD	EXTINCT
NE	DD	LC	NT	VU	EN	CR	EW	EX

© LOWVELD RHINO TRUST, ZIMBABWE

The **Black Rhino**, Diceros bicornis, is listed as 'Critically Endangered' on the IUCN's Red List of Threatened Species™. Once found throughout much of sub-Saharan Africa, over 98% are now protected in state, private and communal conservation areas in South Africa, Namibia, Kenya, Zimbabwe and Tanzania. There are also small re-established breeding populations in Swaziland, Zambia, Malawi and Botswana.

While it was the most numerous of the world's rhino species into the middle of the 20th century, it soon lost this 'status' as numbers plunged due to intensive targeted poaching for its horn. From numbering in the region of 100,000 in the early 1960s, only about 2,400 individuals of the four subspecies of Black Rhino remained by 1995. Fortunately, Black Rhino numbers have steadily recovered since then to about 4,200.

The overall recovery has been largely due to their effective protection and the translocation of surplus rhinos to re-establish viable populations elsewhere within their former range. Controls resulting from its listing on CITES Appendix I (the global convention to address species threatened by international trade) have been an additional positive influence.

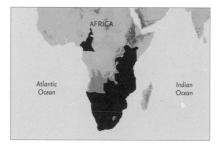

A species is **Critically Endangered** when it faces an extremely high risk of extinction in the wild, based on measurements of population size and/or geographic range and their trends in the past, present and/or future.

NOT EVALUATED	DATA DEFICIENT		LEAST CONCERN	NEAR THREATENED	VULNERABLE	ENDANGERED	CRITICALLY ENDANGERED	EXTINCT IN THE WILD	EXTINCT
NE	DD		LC	NT	VU	EN	CR	EW	EX

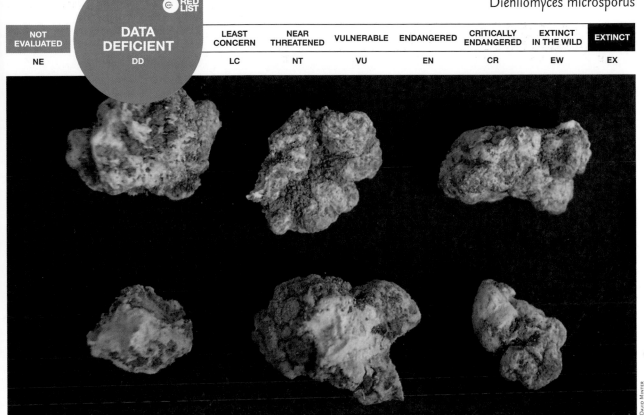

© David Minter

Diehliomyces microsporus has not yet been officially evaluated for the IUCN Red List of Threatened Species™, however it has provisionally been listed as 'Data Deficient'. It is a microscopic fungus, known only from compost used for commercial farming of mushrooms. Claims that it originates from soil, while plausible, lack supporting evidence – it seems never to have been found in any natural habitat.

Mushroom farmers regard Diehliomyces microsporus as a 'fungus weed' because infestations can cause serious economic loss. In recent years, various treatments have been developed, including compost pasteurization. These control the fungus very effectively, and Diehliomyces microsporus infestations have been totally eliminated on farms using the treatments.

The only habitat in which this fungus has ever been observed is thus being made unavailable to it. This species is threatened in cultivation and unknown in the wild, a very unusual situation which raises interesting and problematic conservation issues. Efforts are needed to find its natural habitat.

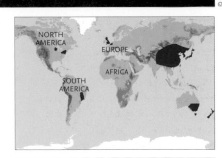

A species is **Data Deficient** when there is inadequate information to make an assessment of its risk of extinction. Listing of species in this category indicates that more information is required, and acknowledges that future research may show that threatened classification is appropriate.

Wandering Albatross

NOT EVALUATED	DATA DEFICIENT	LEAST CONCERN	NEAR THREATENED	⟨ VULNERABLE ⟩	ENDANGERED	CRITICALLY ENDANGERED	EXTINCT IN THE WILD	EXTINCT
NE	DD	LC	NT	VU	EN	CR	EW	EX

A species is **Vulnerable** when it faces a high risk of extinction in the wild, based on measurements of population size and/or geographic range and their trends in the past, present and/or future.

The **Wandering Albatross**, *Diomedea exulans*, is classified as 'Vulnerable' on the IUCN Red List of Threatened Species™. It is one of the largest birds in the world with a wingspan of 2.5–3.35 meters. They spend a large amount of time in flight, soaring over the southern oceans – one bird was recorded to have travelled 6000 km in just 12 days.

The main threat to the Wandering Albatross is incidental mortality when birds are caught on fishing hooks set on longlines by boats in the tuna and Patagonian toothfish industries. Individuals get trapped on the hooks, whilst trying to steal fish and squid bait, and often drown when snared and pulled underwater.

The majority of Wandering Albatross breeding sites are protected within reserves. Adoption of mitigation measures in some areas, coupled with the relocation of other fisheries away from foraging grounds, are very positive steps for conservation but have yet to lead to signs of recovery in most populations. Widespread adoption of mitigation measures in longline fishing practices and ongoing monitoring will be required in order to safeguard the future of these albatrosses.

NOT EVALUATED	DATA DEFICIENT	LEAST CONCERN	NEAR THREATENED	VULNERABLE	ENDANGERED	CRITICALLY ENDANGERED	EXTINCT IN THE WILD	EXTINCT
NE	DD	LC	NT	VU	EN	CR	EW	EX

Dioon caputoi is listed as 'Critically Endangered' on the IUCN Red List of Threatened Species™. There are only four known populations of this small, distinctive cycad, all of which occur within the Tehuacán-Cuicatlán Biosphere Reserve along the border between the Mexican states of Puebla and Oaxaca.

Habitat destruction due to ranching and agriculture has devastated many Mexican cycads. Although plant collecting from the wild has been less severe than elsewhere in the world, various Mexican species, including Dioon caputoi, have been decimated by collectors. None of the populations of this species now number more than 120 individuals, and all comprise mostly adult plants due to poor natural recruitment of seedlings.

Although Dioon caputoi receives some protection from international trade under its listing on Appendix II of CITES, additional conservation measures are still urgently needed to prevent extraction of plants and further habitat loss. Continued population monitoring as part of a long-term management plan, and the additional reintroduction of artificially propagated plants, will contribute to the survival of this range-restricted species.

© JEFF CHEMNICK

A species is Critically Endangered when it faces an extremely high risk of extinction in the wild, based on measurements of population size and/or geographic range and their trends in the past, present and/or future.

Wild Yam

NOT EVALUATED	DATA DEFICIENT	LEAST CONCERN	NEAR THREATENED	VULNERABLE	ENDANGERED	CRITICALLY ENDANGERED	EXTINCT IN THE WILD	EXTINCT
NE	DD	LC	NT	VU	EN	CR	EW	EX

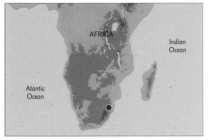

A species is **Critically Endangered** when it faces an extremely high risk of extinction in the wild, based on measurements of population size and/or geographic range and their trends in the past, present and/or future.

The **Wild Yam**, Dioscorea strydomiana, is listed as 'Critically Endangered' on the Red List of South African Plants. While surveying plant material sold at a traditional medicine market, Gerhard Strydom, after whom this plant is named, came across a yam completely different from other known species. To seek out its location, traditional healers led him to a remote valley in the mountains bordering South Africa and Swaziland.

This new species, which is being described by yam experts at the Royal Botanic Gardens, Kew, has a shrub-like growth form, with large woody tubers that grow above the ground, unlike other yams. The Wild Yam is threatened by harvesting of its tubers for traditional medicine. Annual monitoring of two small populations revealed that many tubers are cut often and repeatedly, and that this harvesting is likely to be unsustainable.

Conservation measures, implemented with the cooperation of the tribal authority on whose land the Wild Yam occurs, include control of access to the site, ex situ propagation (the preservation of species outside of their natural habitat) of plants for medicinal and horticultural trade, and banking of seeds with Kew's Millennium Seed Bank Project.

Giant Kangaroo Rat

NOT EVALUATED	DATA DEFICIENT	LEAST CONCERN	NEAR THREATENED	VULNERABLE	ENDANGERED	CRITICALLY ENDANGERED	EXTINCT IN THE WILD	EXTINCT
NE	DD	LC	NT	VU	EN	CR	EW	EX

The **Giant Kangaroo Rat**, Dipodomys ingens, is listed as 'Critically Endangered' on the IUCN Red List of Threatened Species™. This species is endemic to California in the United States, where it occurs in highly fragmented populations along the southwestern border of the San Joaquin Valley, and in a few nearby valleys to the west.

The Giant Kangaroo Rat population plummeted during the 20th century, mainly as a result of habitat loss as desert areas were converted for agricultural uses. Although the conversion of habitat for agriculture has slowed, only around two percent of this species' historical habitat remains, and urban and industrial development pose an ongoing threat to these remaining areas.

This species is listed as Endangered by the California Fish and Game Commission and the United States Fish and Wildlife Service. A recovery plan has been developed in an effort to secure the survival of this species, and populations are protected within the Carrizo Plain Natural Heritage Reserve and a number of Federal lands.

© JOHN ROSER

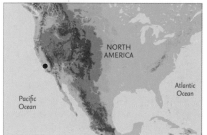

A species is Critically Endangered when it faces an extremely high risk of extinction in the wild, based on measurements of population size and/or geographic range and their trends in the past, present and/or future.

NOT EVALUATED	DATA DEFICIENT	LEAST CONCERN	NEAR THREATENED	VULNERABLE	ENDANGERED	CRITICALLY ENDANGERED	EXTINCT IN THE WILD	EXTINCT
NE	DD	LC	NT	VU	EN	CR	EW	EX

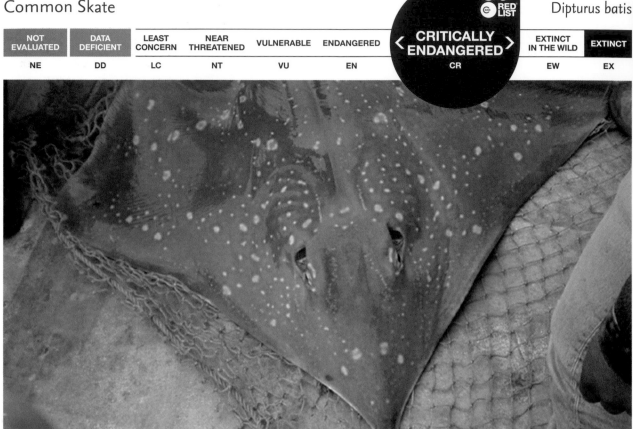

The **Common Skate**, *Dipturus batis*, is listed as 'Critically Endangered' on the IUCN Red List of Threatened Species™. The largest European ray, growing to at least two metres in length, it is found in the northeastern Atlantic Ocean and the Mediterranean Sea.

The Common Skate was once one of the most abundant rays in the northeast Atlantic and made up a large part of commercial catches. Since then, it has undergone dramatic declines, particularly around the British Isles. Caught both by fisheries targeting this species, and as by-catch, its long lifespan and slow maturation mean the Common Skate has little ability to withstand high levels of exploitation. This is probably the most threatened marine fish in the world.

The Common Skate is not subject to any species-specific conservation measures across its range, though it would likely benefit from more general regulations prohibiting fishing in the North Sea and Mediterranean. Suitable non-trawling areas may also be needed to protect both adults and eggs from capture by trawling gear.

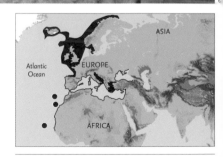

A species is **Critically Endangered** when it faces an extremely high risk of extinction in the wild, based on measurements of population size and/or geographic range and their trends in the past, present and/or future.

Barndoor Skate

Dipturus laevis

© Phillip Blackmon

The **Barndoor Skate**, *Dipturus laevis*, is listed as 'Endangered' on the IUCN Red List of Threatened Species™. As the largest skate in the Northwest Atlantic it is intrinsically sensitive to overexploitation, and it is found in relatively cool waters along the Atlantic coast of the USA and Canada, down to depths of 1,300 m.

The Barndoor Skate was often taken as bycatch in commercial trawl nets, and as part of dogfish and skate fisheries in some areas. Catch rates declined by up to 99 percent between the 1960s and 1990s, leading to warnings that the species was on the brink of extinction. This finding is not without controversy, and numbers may now even be increasing as a result of systemic fisheries being rebuilt in this region.

A number of petitions have been made to list the Barndoor Skate under the US Endangered Species Act, but these have so far been rejected. There has been a ban on possession of the species in US waters, and 'no-take' zones such as that at Georges Bank appear to have allowed the population to start recovering.

A species is **Endangered** when it faces a very high risk of extinction in the wild, based on measurements of population size and/or geographic range and their trends in the past, present and/or future.

Dragon's Blood Tree

Dracaena cinnabari

NOT EVALUATED	DATA DEFICIENT	LEAST CONCERN	NEAR THREATENED	⟨ VULNERABLE ⟩	ENDANGERED	CRITICALLY ENDANGERED	EXTINCT IN THE WILD	EXTINCT
NE	DD	LC	NT	VU	EN	CR	EW	EX

The **Dragon's Blood Tree**, *Dracaena cinnabari*, is listed as 'Vulnerable' on the IUCN Red List of Threatened Species™. Endemic to the island of Socotra, this distinctive tree is famed for its dark red resin, or 'dragon's blood', a substance highly prized since ancient times as a colourant and medicine.

Although the vegetation of Socotra remains relatively well-preserved, it faces growing threats from population expansion, increasing development and overgrazing. The range of the Dragon's Blood Tree has become reduced and fragmented, and its populations are undergoing poor regeneration. Although human activities may have contributed to this decline, the main threat is believed to be the increasing aridity on the island, a process that may be exacerbated by climate change.

Several initiatives are underway to support conservation on Socotra, and the Dragon's Blood Tree is considered to be an important flagship species. However, if this unique plant is to survive, a range of specific measures will be needed, including monitoring of natural regeneration, planting of seedlings, fencing against livestock, and the protection of its habitat.

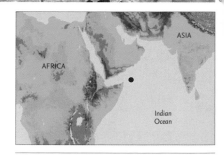

A species is Vulnerable when it faces a high risk of extinction in the wild, based on measurements of population size and/or geographic range and their trends in the past, present and/or future.

Canary Island Dragon Tree

Dracaena draco

© Helena Marathefti

A species is **Vulnerable** when it faces a high risk of extinction in the wild, based on measurements of population size and/or geographic range and their trends in the past, present and/or future.

The **Canary Island Dragon Tree**, *Dracaena draco*, is listed as 'Vulnerable' on IUCN's Red List of Threatened Species™. It is found on the Canary Islands, Cape Verde Islands, Madeira and Morocco.

There has been a recent decline in naturally occurring Dragon Trees and the species is becoming very rare in the wild. It is thought that a now extinct endemic flightless bird that fed on the fruit of the tree may have played an essential part in stimulating germination. This is one theory, however the tree is also threatened by habitat loss and grazing pressure. The tree's red sap, known as 'Dragon's Blood', has a variety of uses as a colourant and medicine, and was used by the Guanche people of the Canary Islands in their mummification processes.

The species is protected under national law from picking and uprooting, and efforts have been made on the islands to make people aware of the plant and the need for its conservation. Outside of its natural habitat, there are 127 collections containing this species.

Lord Howe Island Stick Insect

NOT EVALUATED	DATA DEFICIENT	LEAST CONCERN	NEAR THREATENED	VULNERABLE	ENDANGERED	CRITICALLY ENDANGERED	EXTINCT IN THE WILD	EXTINCT
NE	DD	LC	NT	VU	EN	CR	EW	EX

The **Lord Howe Island Stick Insect**, *Dryococelus australis*, is listed as 'Critically Endangered' on the IUCN Red List of Threatened Species™. It is known only from Lord Howe Island and Ball's Pyramid, a volcanic outcrop in the Tasman Sea just 200 m wide at the base. Young nymphs are bright green in colour and become darker as they grow, eventually turning a dark glossy brown or even black. Females are larger than males and can reach 13 cm in length.

The introduction of predatory black rats to the island by the trading vessel SS Makambo in 1918 led to the extinction of the species on Lord Howe Island, possibly as early as 1920. The Lord Howe Island Stick Insect was considered extinct in 1986; however, a small number of this species had survived on Ball's Pyramid. Following research in the early 2000s, a pair was taken to begin a captive breeding programme.

This species is being reared successfully in captivity, and there are plans for a reintroduction to Lord Howe Island if the eradication of the rats is successful.

A species is **Critically Endangered** when it faces an extremely high risk of extinction in the wild, based on measurements of population size and/or geographic range and their trends in the past, present and/or future.

Pemba Palm

Dypsis pembana

© Antje Förstle

The **Pemba Palm**, Dypsis pembana, is listed as 'Vulnerable' on the IUCN Red List of Threatened Species™. This attractive palm is found only on the tiny Tanzanian island of Pemba, off the east coast of Africa.

Owing to a relatively large human population, there is considerable pressure on natural resources on Pemba, and a large proportion of the island's forest has been destroyed or severely degraded. The primary threats to the remaining forest include fires, logging and exploitation, agricultural conversion, the spread of invasive plants, and the expansion of tourist infrastructure.

The **Global Trees Campaign** is involved in a project that began in early 2009 with the aim of conserving the Pemba Palm. The main priorities include the implementation of further surveys and research, the management of fires, the removal of invasive plants, and awareness-raising amongst the local community. In addition, plans are in place to develop tree nurseries, which will provide seedlings to reinforce the wild population, whilst also giving local people the opportunity to grow fruit and timber trees.

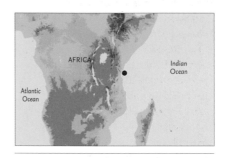

A species is Vulnerable when it faces a high risk of extinction in the wild, based on measurements of population size and/or geographic range and their trends in the past, present and/or future.

NOT EVALUATED	DATA DEFICIENT	LEAST CONCERN	NEAR THREATENED	⟨ VULNERABLE ⟩	ENDANGERED	CRITICALLY ENDANGERED	EXTINCT IN THE WILD	EXTINCT
NE	DD	LC	NT	VU	EN	CR	EW	EX

© César García

Echinodium renauldii is listed as 'Vulnerable' on the IUCN Red List of Threatened Species™. This rare moss is only found in the Azores (Portugal), where it is known from five different islands. It grows on rocks in forested, deeply shaded ravines and craters.

Because of changes in Portuguese land policy, the laurel forest habitat in which *Echinodium renauldii* occurs is becoming more and more threatened by logging. In addition, patches of forest are still being converted for pasture, while livestock activity is leading to water nitrification which is locally threatening some populations. Most recently, the species has been increasingly affected by the aggressive invasive plant *Hedychium gardnerianum*, introduced from India.

In the absence of any specific conservation measures for *Echinodium renauldii*, this species is likely to decline further and may need elevating to a more threatened Red List category in the future.

A species is **Vulnerable** when it faces a **high risk** of extinction in the wild, based on measurements of population size and/or geographic range and their trends in the past, present and/or future.

Rabb's Fringe-limbed Treefrog

Ecnomiohyla rabborum

NOT EVALUATED	DATA DEFICIENT	LEAST CONCERN	NEAR THREATENED	VULNERABLE	ENDANGERED	CRITICALLY ENDANGERED	EXTINCT IN THE WILD	EXTINCT
NE	DD	LC	NT	VU	EN	CR	EW	EX

© Brad Wilson

Rabb's Fringe-limbed Treefrog, Ecnomiohyla rabborum, is listed as 'Critically Endangered' on the IUCN Red List of Threatened Species™. This recently discovered species is found only in the mountains near El Valle de Anton, in central Panama. It inhabits the forest canopy, and may leap from the trees to escape danger, using its outstretched limbs and large, webbed hands and feet to glide safely to the ground.

Already uncommon at the time of its discovery, this rare amphibian may possibly have become extinct in the wild after the fungal disease chytridiomycosis arrived in the region in 2006. In addition to disease, forest clearance for the development of luxury holiday homes threatens the species' habitat, and its restriction to a single, small site makes any remaining population particularly vulnerable.

A captive breeding programme is underway for Rabb's Fringe-limbed Treefrog, but has yet to produce positive results. Field surveys are continuing in an attempt to locate any surviving wild individuals, but with perhaps just a single female now remaining, the species' situation appears to be critical.

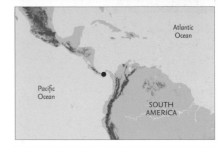

A species is Critically Endangered when it faces an extremely high risk of extinction in the wild, based on measurements of population size and/or geographic range and their trends in the past, present and/or future.

Ectophylla alba

NOT EVALUATED	DATA DEFICIENT	LEAST CONCERN	NEAR THREATENED	VULNERABLE	ENDANGERED	CRITICALLY ENDANGERED	EXTINCT IN THE WILD	EXTINCT
NE	DD	LC	NT	VU	EN	CR	EW	EX

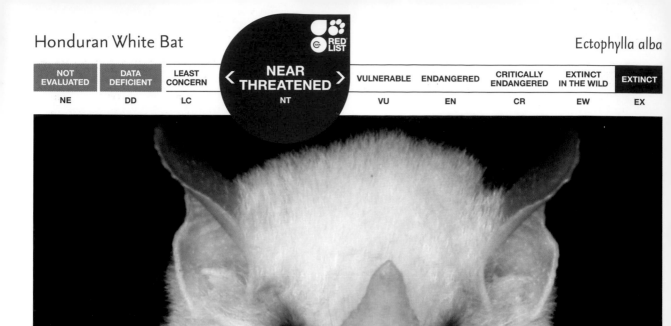

The **Honduran White Bat**, *Ectophylla alba*, is listed as 'Near Threatened' on the IUCN Red List of Threatened Species™. True to its name, this tiny bat has a distinctive white coat, and is restricted to specific habitats in the lowlands of the Caribbean slopes of Honduras, Nicaragua, Costa Rica, and northwestern Panama. It feeds on the fruits of very few species of trees, and is one of a number of bat species that roost in 'tents' formed by altering the shape of plant leaves.

As a rainforest inhabitant, the loss and degradation of Central America's forests is the principal threat to the Honduran White Bat. The population is currently undergoing a significant decline, and in the absence of conservation measures is likely to be up-listed to a more threatened category in the future.

At present, there are no specific conservation measures in place for the Honduran White Bat, but it is at least known to occur in several protected areas.

A species is Near Threatened when it does not qualify for Critically Endangered, Endangered or Vulnerable now, but is close to qualifying for or is likely to qualify for a threatened category in the near future.

Asian Elephant

Elephas maximus

NOT EVALUATED	DATA DEFICIENT	LEAST CONCERN	NEAR THREATENED	VULNERABLE	‹ ENDANGERED ›	CRITICALLY ENDANGERED	EXTINCT IN THE WILD	EXTINCT
NE	DD	LC	NT	VU	EN	CR	EW	EX

© Ajay A Desai

The **Asian Elephant**, *Elephas maximus*, is listed as 'Endangered' on the IUCN Red List of Threatened Species™. It is found in isolated populations in 13 tropical Asian countries. The Asian Elephant is smaller than its African savannah relative; the ears are smaller and the back is more rounded.

The numbers of Asian Elephants have been decimated by habitat loss, degradation, and fragmentation, driven by an expanding human population. This causes elephants to become increasingly isolated, often coming into conflict with local farmers. Crops are damaged and lives lost; up to 300 people a year are killed by elephants in India. Poaching for ivory is also a threat and because only males have tusks, populations can become extremely skewed towards females, thus affecting breeding rates.

The most important conservation priorities for the Asian Elephant are: conservation of their habitat and maintaining habitat connectivity by securing corridors; management of human- elephant conflicts; improved legislation and law enforcement with enhanced field patrolling; and regulating/curbing trade in ivory and other elephant products.

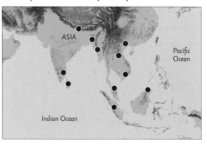

Venda Cycad

NOT EVALUATED	DATA DEFICIENT	LEAST CONCERN	NEAR THREATENED	VULNERABLE	ENDANGERED	‹ CRITICALLY ENDANGERED ›	EXTINCT IN THE WILD	EXTINCT
NE	DD	LC	NT	VU	EN	CR	EW	EX

The **Venda Cycad**, Encephalartos hirsutus, is listed as 'Critically Endangered' on the IUCN Red List of Threatened Species™. Described as a new species in 1996, this cycad was originally known from small populations in three localities in the Limpopo Province of South Africa. The species name hirsutus, which means 'hairy', is a reference to the dense covering of hairs that are found on the emergent leaves.

The species has declined dramatically in the past decade due to illegal collecting, and unconfirmed reports suggest that the Venda Cycad may be 'Extinct in the Wild'. Over the past few decades, many South African cycads have become increasingly scarce in the wild, mainly as a result of illegal collecting. Bark harvesting for traditional medicine, habitat loss, and the spread of alien vegetation are additional threats to cycad populations.

There is no formal protection of the Venda Cycad habitat, but this species is listed in South Africa's Threatened or Protected Species regulations and on Appendix I of CITES (which makes it illegal to trade this species internationally). A reintroduction programme using confiscated plants as a seed source is also being considered.

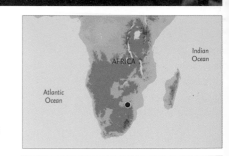

A species is Critically Endangered when it faces an extremely high risk of extinction in the wild, based on measurements of population size and/or geographic range and their trends in the past, present and/or future.

Albany Cycad

NOT EVALUATED	DATA DEFICIENT	LEAST CONCERN	NEAR THREATENED	VULNERABLE	ENDANGERED	CRITICALLY ENDANGERED	EXTINCT IN THE WILD	EXTINCT
NE	DD	LC	NT	VU	EN	CR	EW	EX

The **Albany Cycad**, Encephalartos latifrons, is listed as 'Critically Endangered' on the IUCN Red List of Threatened Species™. It is a slow-growing cycad that occurs in a small area of the Eastern Cape Province in South Africa. No natural reproduction has occurred for more than 30 years, probably due to the distance between plants and the local extinction of beetle pollinators.

The Albany Cycad probably first declined due to habitat loss, but in the last 50 years the prevailing threat has been the removal of adult plants by plant collectors. This is a common problem for all southern African cycads. Invasive plants, especially Hakea sericia, have transformed the habitat in some areas and could become an additional threat.

None of the surviving subpopulations of Albany Cycad occur naturally within protected areas, but a species management plan has been developed to address the main threats. The plan includes actions to improve the security of plants on private land, artificial pollination and propagation of seeds, and reintroduction and translocation of plants to protected sites.

© John Donaldson

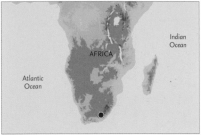

A species is **Critically Endangered** when it faces an **extremely high risk of extinction** in the wild, based on measurements of population size and/or geographic range and their trends in the past, present and/or future.

AFRICA

Indian Ocean

Albany Cycad • Encephalartos latifrons   159

NOT EVALUATED	DATA DEFICIENT	LEAST CONCERN	NEAR THREATENED	VULNERABLE	ENDANGERED	CRITICALLY ENDANGERED	EXTINCT IN THE WILD	EXTINCT
NE	DD	LC	NT	VU	EN	CR	EW	EX

Encephalartos whitelockii is listed as 'Critically Endangered' on the IUCN Red List of Threatened Species™. This spectacular cycad is endemic to the Mpanga Gorge in Uganda, where it forms one of the most impressive cycad populations in Africa.

The greatest threat to Encephalartos whitelockii is the ongoing construction of a hydroelectric power station above the Mpanga Falls. The development of this new power station resulted in the elevation of this cycad's IUCN Red List status, as a large proportion of the population is affected by the construction. The activities of local communities pose an additional threat, with seasonal fires and deliberate felling being particularly problematic.

Owing to the increased pressure on the Encephalartos whitelockii population, mitigation measures for the site include a local community-based conservation project, which aims to raise community awareness of the conservation and importance of the species, as well as a seed bank and nursery. The propagation of this species in a nursery will provide individuals for replanting in the wild.

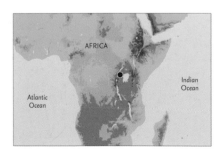

Chapman's Blenny

Entomacrodus chapmani

RED LIST

NOT EVALUATED	DATA DEFICIENT	LEAST CONCERN	NEAR THREATENED	‹ VULNERABLE ›	ENDANGERED	CRITICALLY ENDANGERED	EXTINCT IN THE WILD	EXTINCT
NE	DD	LC	NT	VU	EN	CR	EW	EX

© Dr. John E. Randall

Chapman's Blenny, *Entomacrodus chapmani*, is classified as 'Vulnerable' on the IUCN Red List of Threatened Species™. This species of blenny is found only in shallow, rocky waters (less than 2 metres) and tide-pools surrounding Easter Island. It is estimated that this species occupies an area of habitat no greater than 1 km².

While this species is not sought after as food or for the aquarium trade, it is still considered 'Vulnerable' because of its extremely limited range and the potential for future reduction in the quality of its habitat. Over the past 10 years, the number of tourists travelling to Easter Island has more than doubled, and is expected to increase further in the coming years. As tourism on the island continues to grow, so will the amount of coastal development, leading to further loss of suitable habitat for Chapman's Blenny.

Currently there is no protective legislation in place for this species, but some of its range does fall within three Marine Protected Areas (MPAs) around Easter Island. However, this area alone cannot ensure the survival of this species.

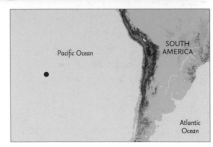

A species is Vulnerable when it faces a high risk of extinction in the wild, based on measurements of population size and/or geographic range and their trends in the past, present and/or future.

Goliath Grouper

NOT EVALUATED	DATA DEFICIENT	LEAST CONCERN	NEAR THREATENED	VULNERABLE	ENDANGERED	CRITICALLY ENDANGERED	EXTINCT IN THE WILD	EXTINCT
NE	DD	LC	NT	VU	EN	CR	EW	EX

The **Goliath Grouper**, Epinephelus itajara, is listed as 'Critically Endangered' on the IUCN Red List of Threatened Species™. As its name suggests, this large, stocky grouper is the second largest member of the grouper family. It is found in tropical and subtropical waters of the Atlantic Ocean, including the Gulf of Mexico and the Caribbean Sea.

The Goliath Grouper is fished both commercially and for sport, but its slow growth and reproductive rates, and its group spawning behaviour, make it particularly vulnerable to overfishing. Since the 1970s, catch rates and average sizes have declined sharply, with the loss of critical mangrove habitat also affecting the recruitment of juveniles into the population.

A number of regulations are now in place to prohibit the harvesting of Goliath Groupers. The species is showing promising signs of recovery in some areas, but continued surveys, education programmes and the species' inclusion in Marine Protected Areas have been recommended, and many more years of protection will be needed if its populations are to properly recover.

A species is Critically Endangered when it faces an extremely high risk of extinction in the wild, based on measurements of population size and/or geographic range and their trends in the past, present and/or future.

Nassau Grouper

Epinephelus striatus

© SHUTTERSTOCK - TUBUCEO

The **Nassau Grouper**, *Epinephelus striatus*, is listed as 'Endangered' on the IUCN Red List of Threatened Species™. This large reef fish is found in shallow waters of the tropical western Atlantic Ocean and the Caribbean Sea.

Historically, the Nassau Grouper was one of the most important commercially harvested fishes throughout the region, with its habit of gathering in huge numbers in predictably located spawning (reproductive) aggregations making it an easy target. However, over the last few decades, the species has undergone a significant decline throughout its range, and is now commercially extinct in a number of areas.

Following massive lobbying by conservationists, and scientific studies, the Nassau Grouper has received some of the protection needed if numbers are to recover from past exploitation. Fishing for Nassau Grouper is banned in US Federal waters, including Puerto Rico and the US Virgin Islands, spawning sites are protected in the Cayman Islands, and fishing is banned in Belize, Mexico, and the Dominican Republic during the spawning season.

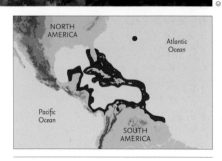

A species is Endangered when it faces a very high risk of extinction in the wild, based on measurements of population size and/or geographic range and their trends in the past, present and/or future.

Oyster Mussel

NOT EVALUATED	DATA DEFICIENT	LEAST CONCERN	NEAR THREATENED	VULNERABLE	ENDANGERED	CRITICALLY ENDANGERED	EXTINCT IN THE WILD	EXTINCT
NE	DD	LC	NT	VU	EN	CR	EW	EX

© Arthur E. Bogan, North Carolina State Museum of Natural Sciences

A species is Critically Endangered when it faces an extremely high risk of extinction in the wild, based on measurements of population size and/or geographic range and their trends in the past, present and/or future.

The **Oyster Mussel**, *Epioblasma capsaeformis*, is listed as 'Critically Endangered' on the IUCN Red List of Threatened Species™. Once widespread across four southeastern US states, this freshwater mussel is now found only in a small number of streams and rivers in Alabama, Kentucky, Tennessee and Virginia.

Human modifications to rivers and streams, including channel modifications, pollution, and water impoundments, have had a serious impact on this species and its habitat. In addition to affecting water flow and quality, they have produced barriers to dispersal, and by the year 2000 the Oyster Mussel had undergone an estimated 80% population decline. The small, isolated populations that remain are also under increased risk of being wiped out by isolated events such as chemical spills.

A recovery plan in place for the Oyster Mussel recommends protecting remaining populations and habitat, establishing additional populations, improving habitat quality, and undertaking further research into the species. Techniques are also being developed for artificial rearing, with the ultimate aim of reintroducing the species into parts of its historical range.

African Wild Ass

Equus africanus

NOT EVALUATED	DATA DEFICIENT	LEAST CONCERN	NEAR THREATENED	VULNERABLE	ENDANGERED	CRITICALLY ENDANGERED	EXTINCT IN THE WILD	EXTINCT
NE	DD	LC	NT	VU	EN	CR	EW	EX

© Patricia D. Moehlman 2005

The **African Wild Ass**, *Equus africanus*, is listed as 'Critically Endangered' on the IUCN Red List of Threatened Species™. The ancestor of the domestic donkey, this species once ranged across much of northern Africa, but now only occurs in scattered populations in Eritrea, Ethiopia, and Somalia, and possibly also in Djibouti, Sudan and Egypt.

The greatest threat to the African Wild Ass is hunting for food and traditional medicine. There are now estimated to be fewer than 200 mature individuals remaining in the wild, with no subpopulation numbering more than 50 mature individuals. Limited access to drinking water and forage is a major constraint to reproduction and survival in a desert habitat that has severe and recurrent droughts.

Although the African Wild Ass is legally protected in the countries where it is currently found, more effective protection measures need to be adopted. In Eritrea and Ethiopia, the involvement of local pastoralists in research and conservation programmes has been vital in sustaining African Wild Ass populations.

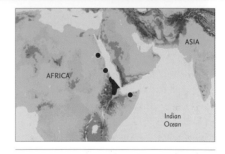

A species is Critically Endangered when it faces an extremely high risk of extinction in the wild, based on measurements of population size and/or geographic range and their trends in the past, present and/or future.

Przewalski's Horse

NOT EVALUATED	DATA DEFICIENT	LEAST CONCERN	NEAR THREATENED	VULNERABLE	ENDANGERED	CRITICALLY ENDANGERED	EXTINCT IN THE WILD	EXTINCT
NE	DD	LC	NT	VU	EN	CR	EW	EX

RED LIST

© Patricia D Moehlman

Przewalski's Horse, Equus ferus przewalskii, is listed as 'Critically Endangered' on the IUCN Red List of Threatened Species™. Przewalski's Horse is now the last true species of wild horse, and in 1969 the last wild individual was recorded in southwest Mongolia. Previously classified as 'Extinct in the Wild', the release of captive-bred individuals and the survival of their offspring in the wild to maturity led to it being down-listed to 'Critically Endangered'.

Hunting, human conflict, competition with domestic livestock, habitat degradation and capture expeditions were all thought to have caused the extinction of Przewalski's Horse in the wild. Today, reintroduced populations are primarily threatened by severe winters, resource exploitation, and contact with domestic horses which can lead to hybridization and transfer of disease.

Przewalski's Horse is legally protected in Mongolia where reintroductions began in the 1990s. There are now three reintroduction sites, including Hustai National Park, Great Gobi B Strictly Protected Area, and Seriin Nuruu. For the past four years there have been more than 50 mature Przewalski's Horses surviving in the wild. The re-introduction of this species, and the down-listing of its threatened status, is a true success story for conservation.

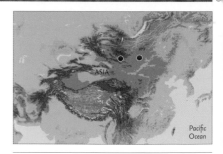

A species is **Critically Endangered** when it faces an extremely high risk of extinction in the wild, based on measurements of population size and/or geographic range and their trends in the past, present and/or future.

Grevy's Zebra

Equus grevyi

NOT EVALUATED	DATA DEFICIENT	LEAST CONCERN	NEAR THREATENED	VULNERABLE	‹ ENDANGERED ›	CRITICALLY ENDANGERED	EXTINCT IN THE WILD	EXTINCT
NE	DD	LC	NT	VU	EN	CR	EW	EX

© Patricia D Moehlman, 2004

Grevy's Zebra, Equus grevyi, is listed as 'Endangered' on the IUCN Red List of Threatened Species™. It is restricted to Ethiopia and Kenya, and is most easily distinguished from the more common Plains Zebra by its narrower stripes.

Grevy's Zebra has declined by over 50 percent in recent decades, with only around 750 mature individuals now left in the wild. Hunting for skins in the 1970s may have contributed to its decline, but habitat loss through overgrazing, competition with livestock, and a reduced water supply are now the major threats. In some areas hunting continues to be a major threat, while disease outbreaks pose an additional hazard to the already reduced population.

Grevy's Zebra is legally protected in Ethiopia and safeguarded by a hunting ban in Kenya. At present, protected areas only cover a small portion of its range, although they may prove to be crucial to the preservation of some populations. Kenya has developed a national conservation strategy for the species, and research and community-based conservation efforts are ongoing in both Kenya and Ethiopia.

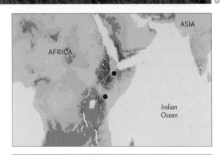

A species is **Endangered** when it faces a very high risk of extinction in the wild, based on measurements of population size and/or geographic range and their trends in the past, present and/or future.

Hawksbill Turtle

Eretmochelys imbricata

NOT EVALUATED	DATA DEFICIENT	LEAST CONCERN	NEAR THREATENED	VULNERABLE	ENDANGERED	CRITICALLY ENDANGERED	EXTINCT IN THE WILD	EXTINCT
NE	DD	LC	NT	VU	EN	CR	EW	EX

© Nicolas Pilcher

The **Hawksbill Turtle**, Eretmochelys imbricata, is classified as 'Critically Endangered' on the IUCN Red List of Threatened Species™. It is a migratory marine reptile, found in tropical and, to a lesser extent, subtropical waters.

Hawksbill Turtles face a variety of hazards from: 1) Direct take – for food or for the tortoiseshell trade; 2) Fishing impacts – incidental capture and entanglement in nets and long lines ; 3) Beachfront development – that alters or destroys nesting beach habitat; 4) Pollution – ingestible plastics in the ocean; and 5) Climate change – this may affect breeding, since sea turtle sex ratios are dependent upon the temperature of incubation. Hawksbills are also acutely threatened by loss of coral reef communities which act as their feeding sites.

Numerous countries have temporarily or permanently banned all exploitation of sea turtles and their eggs, and are attempting to improve enforcement of international bans on the tortoiseshell trade (although extensive illegal trafficking still occurs). Preventing this black market trade, increasing public awareness, and protecting nesting and foraging areas, are key to the protection of Hawksbills.

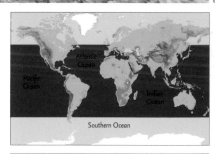

A species is Critically Endangered when it faces an extremely high risk of extinction in the wild, based on measurements of population size and/or geographic range and their trends in the past, present and/or future.

Erica verticillata

NOT EVALUATED	DATA DEFICIENT	LEAST CONCERN	NEAR THREATENED	VULNERABLE	ENDANGERED	CRITICALLY ENDANGERED	EXTINCT IN THE WILD	EXTINCT
NE	DD	LC	NT	VU	EN	CR	EW	EX

Erica verticillata is listed as 'Extinct in the Wild' on the Red List of South African Plants. It disappeared from its native habitat unnoticed sometime during the early 20th century, and was believed to be extinct until cultivated plants were discovered surviving in botanical gardens in South Africa and Europe.

This species once grew profusely on the Cape Flats, a botanical hotspot. However, agricultural and urban expansion, and alien plant invasions, have led to extensive loss of natural vegetation and the local extinction of many plant species such as Erica verticillata. As a result, the Cape Flats now contain South Africa's highest concentration of threatened and extinct plants.

After its rediscovery in botanical gardens, where some cultivated plants were maintained for more than two hundred years, clones of Erica verticillata were propagated and reintroduced into two small nature reserves within its former native range.

© PETER SWART

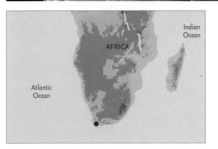

A species is **Extinct in the Wild** when it is known only to survive in cultivation or in captivity. A species is presumed Extinct in the Wild when exhaustive surveys in known and/or expected habitat, at appropriate times throughout its historic range, have failed to record an individual.

Black-breasted Puffleg

NOT EVALUATED	DATA DEFICIENT	LEAST CONCERN	NEAR THREATENED	VULNERABLE	ENDANGERED	CRITICALLY ENDANGERED	EXTINCT IN THE WILD	EXTINCT
NE	DD	LC	NT	VU	EN	CR	EW	EX

The **Black-breasted Puffleg**, Eriocnemis nigrivestis, qualifies as 'Critically Endangered' on the IUCN Red List of Threatened Species™. It is a species of hummingbird known to occur at only two locations in Ecuador.

The main threat to the Black-breasted Puffleg is the felling of forest for timber and charcoal, which facilitates the introduction of cattle for ranching and production of crops. Around 93 percent of the suitable habitat within its probable historic range has been degraded or destroyed. Construction of a pipeline at Cerro Chiquilipe has led to further habitat destruction, and human-induced fires that are started annually in the dry season also threaten remaining areas of forest.

Media coverage of threats to the Black-breasted Puffleg has encouraged the authorities to control access to remaining forest and forbid charcoal production within its key range. Conservation actions include community outreach programmes, environmental education, reforestation on cattle pastures and other degraded lands, the initiation of a monitoring programme, and the formation of a Local Conservation Group to implement these actions.

A species is Critically Endangered when it faces an extremely high risk of extinction in the wild, based on measurements of population size and/or geographic range and their trends in the past, present and/or future.

Boreal Felt Lichen

NOT EVALUATED	DATA DEFICIENT	LEAST CONCERN	NEAR THREATENED	VULNERABLE	ENDANGERED	CRITICALLY ENDANGERED	EXTINCT IN THE WILD	EXTINCT
NE	DD	LC	NT	VU	EN	CR	EW	EX

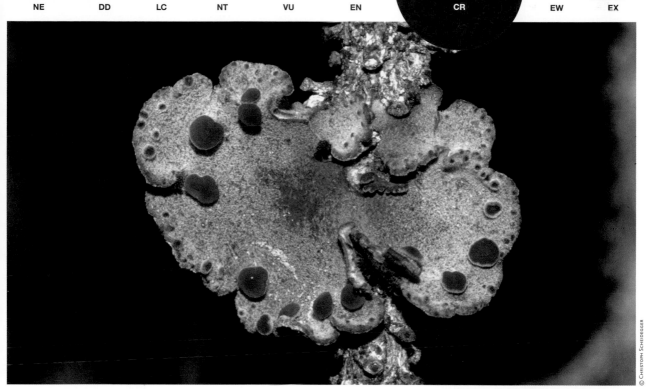

© Christoph Scheidegger

The **Boreal Felt Lichen**, *Erioderma pedicellatum*, is listed as 'Critically Endangered' on the IUCN Red List of Threatened Species™. Originally known from Norway, Sweden and Canada, this rare lichen is now restricted to just one small Alaskan and two Canadian populations, on Newfoundland and Nova Scotia.

The Boreal Felt Lichen has declined by over 80 percent in recent times, mainly as a result of air pollution and habitat loss. In addition to its sensitivity to atmospheric pollutants such as acid rain, this species is threatened by logging

of its forest habitat, which removes the trees on which the lichen grows, and also alters the microclimate of the forest.

The tiny Nova Scotian population of the Boreal Felt Lichen is protected under the Federal Species at Risk Act (SARA), and is the focus of ongoing recovery efforts. Measures are also underway to try and formally protect the species' habitat, and to encourage forest managers to implement management plans to prevent further habitat loss, hopefully giving the Boreal Felt Lichen a brighter future.

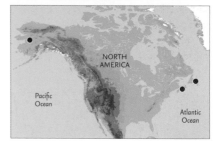

A species is Critically Endangered when it faces an extremely high risk of extinction in the wild, based on measurements of population size and/or geographic range and their trends in the past, present and/or future.

North Atlantic Right Whale

Eubalaena glacialis

NOT EVALUATED	DATA DEFICIENT	LEAST CONCERN	NEAR THREATENED	VULNERABLE	❮ ENDANGERED ❯	CRITICALLY ENDANGERED	EXTINCT IN THE WILD	EXTINCT
NE	DD	LC	NT	VU	EN	CR	EW	EX

The **North Atlantic Right Whale**, *Eubalaena glacialis*, is listed as 'Endangered' on the IUCN Red List of Threatened Species™. Historically, it was common on both sides of the Atlantic; today, this whale appears to be effectively extinct in the eastern North Atlantic.

The North Atlantic Right Whale is currently one of the rarest of the large whales, having been hunted relentlessly for centuries before being given protection in the 1930s. The main threats are ship collisions and entanglements in commercial fishing gear. Other threats include man-made underwater noise (which may interfere with communication and foraging) and habitat loss due to intensive human use of bays and harbours.

Efforts are underway in the US and Canada to limit deaths and injuries due to ship strikes and entanglements. Shipping routes have been changed and ship speed limits introduced around some ports to reduce collision risks. Regulations are also in place in the US requiring modifications to fishing gear and restricting the use of certain types of gear in areas and at times when right whales are common.

A species is Endangered when it faces a very high risk of extinction in the wild, based on measurements of population size and/or geographic range and their trends in the past, present and/or future.

NOT EVALUATED	DATA DEFICIENT	LEAST CONCERN	NEAR THREATENED	‹ VULNERABLE ›	ENDANGERED	CRITICALLY ENDANGERED	EXTINCT IN THE WILD	EXTINCT
NE	DD	LC	NT	VU	EN	CR	EW	EX

© Maria del Carmen Algaba Rubio

Euphorbia cedrorum is listed as 'Vulnerable' on the IUCN Red List of Threatened Species™. Referred to as a 'coralliform' euphorbia because of the crown's resemblance to coral, Euphorbia cedrorum was first described in 1993 from a plant that was being cultivated in a French botanical garden. Later that same year, a small number of mature specimens were discovered at a single location in southwest Madagascar.

The habitat in which the remaining specimens occur is highly threatened and prone to clear-cutting for charcoal production.

Euphorbia cedrorum receives some protection from international trade under its listing on Appendix II of the Convention on International Trade in Endangered Species (CITES), which requires international trade to be carefully controlled through the use of trade permits.

Fortunately, specimens are also still being cultivated in botanical gardens. Nonetheless, in the absence of more direct conservation measures, this species is likely to soon qualify for classification as 'Critically Endangered' or even 'Extinct in the Wild' on the IUCN Red List.

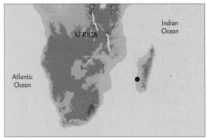

A species is Vulnerable when it faces a high risk of extinction in the wild, based on measurements of population size and/or geographic range and their trends in the past, present and/or future.

NOT EVALUATED	DATA DEFICIENT	LEAST CONCERN	NEAR THREATENED	VULNERABLE	ENDANGERED	CRITICALLY ENDANGERED	EXTINCT IN THE WILD	EXTINCT
NE	DD	LC	NT	VU	EN	CR	EW	EX

© QUENTIN LUKE

Euphorbia tanaensis is listed as 'Critically Endangered' on the IUCN Red List of Threatened Species™, following reassessment in 2006 by the East African Plant Red List Authority. A medium-sized tree, it grows in semi-deciduous swamp forest, and is known from just a single location in Kenya.

In addition to being confined to a single site, the Euphorbia tanaensis population was recently found to number no more than four mature individuals, making it highly vulnerable to extinction. Threats to the species include illegal logging and encroachment of farming activities on the forest boundary.

Euphorbia tanaensis occurs entirely within the Witu Forest Reserve, a protected area of just 42 square kilometres. However, although the area is legally protected, civil insecurity has led to a lack of enforcement and a lack of research, meaning that this highly endangered tree remains under severe threat of extinction. It is known to be in cultivation with two conservationists, and plans are underway to seek funds for multiplication and reintroduction.

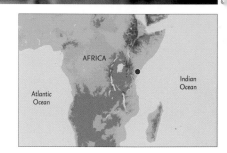

A species is **Critically Endangered** when it faces an extremely high risk of extinction in the wild, based on measurements of population size and/or geographic range and their trends in the past, present and/or future.

Spoon-billed Sandpiper

RED LIST

NOT EVALUATED	DATA DEFICIENT	LEAST CONCERN	NEAR THREATENED	VULNERABLE	ENDANGERED	CRITICALLY ENDANGERED	EXTINCT IN THE WILD	EXTINCT
NE	DD	LC	NT	VU	EN	CR	EW	EX

A species is Critically Endangered when it faces an extremely high risk of extinction in the wild, based on measurements of population size and/or geographic range and their trends in the past, present and/or future.

© ZHENG JIANPING

The **Spoon-billed Sandpiper**, Eurynorhynchus pygmeus, is listed as 'Critically Endangered' on the IUCN Red List of Threatened Species™. It breeds in northeastern Russia and winters in Bangladesh, Myanmar (Burma), Thailand and some other areas of Southeast Asia.

The Spoon-billed Sandpiper is currently experiencing an extremely rapid decline in numbers.

The reason for this change in fortune is not well understood. Habitat changes and disturbance to its breeding grounds may be important, but the loss of staging posts on migration routes are perhaps the biggest difficulty, as South Korea and China have reclaimed huge areas of mudflats for food production. Meanwhile, in the winter in Myanmar, waders are hunted indiscriminately for food.

It is likely that all of these factors need to be addressed by conservationists, although the problem of land reclamation is almost insurmountable. At present, researchers are conducting as many surveys and studies as possible to get a clearer idea of the causes of the decline and what can be done to reverse it.

NOT EVALUATED	DATA DEFICIENT	LEAST CONCERN	NEAR THREATENED	VULNERABLE	ENDANGERED	CRITICALLY ENDANGERED	EXTINCT IN THE WILD	EXTINCT
NE	DD	LC	NT	VU	EN	CR	EW	EX

Fissidens hydropogon is listed as 'Critically Endangered' on the IUCN Red List of Threatened Species™. This robust, much-branched moss grows, periodically submerged, in flowing rivers and streams in rainforest. It was previously known only from the Bombonasa River in Ecuador, but in 2008 was rediscovered in similar habitat along the Rio Nangaritza, 500 km to the southwest.

This rare moss is known only from two tiny areas, and its habitat is believed to be continuing to decline in quality and extent, putting the species at ever-increasing risk of extinction. Its restriction to just two locations also puts it at risk from extreme events, which could wipe out the entire population.

There are not known to be any specific conservation measures in place for Fissidens hydropogon, but it is likely to benefit from any efforts to conserve its rainforest habitat. Further survey work has also been recommended to evaluate the status of this and other highly threatened Ecuadorian bryophytes.

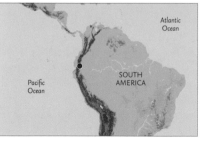

A species is **Critically Endangered** when it faces an extremely high risk of extinction in the wild, based on measurements of population size and/or geographic range and their trends in the past, present and/or future.

Patagonian Cypress

NOT EVALUATED	DATA DEFICIENT	LEAST CONCERN	NEAR THREATENED	VULNERABLE	‹ ENDANGERED ›	CRITICALLY ENDANGERED	EXTINCT IN THE WILD	EXTINCT
NE	DD	LC	NT	VU	EN	CR	EW	EX

A species is Endangered when it faces a very high risk of extinction in the wild, based on measurements of population size and/or geographic range and their trends in the past, present and/or future.

The **Patagonian Cypress**, Fitzroya cupressoides, is listed as 'Endangered' on the IUCN Red List of Threatened Species™. Known locally as 'alerce', this giant conifer occurs in southern Chile and Argentina, and can grow to an impressive 60 metres.

The wood of the Patagonian Cypress is durable, lightweight and highly prized, and the species has been logged since the middle of the 17th Century. Unfortunately, this exploitation occurred at such high intensity that stands of Patagonian Cypress now only occupy around 15 percent of their original area. The species is still illegally logged and its naturally slow regeneration means that any harvest is unsustainable.

Logging of the Patagonian Cypress is prohibited throughout its range, although this is not always adhered to, and illegal logging has been difficult to halt in remote areas. To assist propagation of the species, research is being carried out into the conditions required for the Patagonian Cypress to regenerate, and the Global Trees Campaign is also involved in efforts to protect and conserve the remaining forest areas.

Restinga Antwren

NOT EVALUATED	DATA DEFICIENT	LEAST CONCERN	NEAR THREATENED	VULNERABLE	ENDANGERED	CRITICALLY ENDANGERED	EXTINCT IN THE WILD	EXTINCT
NE	DD	LC	NT	VU	EN	CR	EW	EX

The **Restinga Antwren**, *Formicivora littoralis*, is listed as 'Critically Endangered' on the IUCN Red List of Threatened Species™. Rare and highly specialized, this small antwren is restricted to a unique coastal ecosystem on the coast of Rio de Janeiro state, Brazil.

The restricted habitat on which this species depends ('restinga' is a type of tall, dry, coastal scrub) is under enormous pressure from beachfront development for housing and holiday resorts, and from the salt industry. Owing to a burgeoning human population, squatters also pose a threat to its habitat, while nest predation by alien predators has now become an additional concern.

Fortunately, conservation efforts have proliferated in recent years, with measures including the implementation of an awareness campaign aimed at local schools, the preparation of a species action plan, the removal of alien predators, and the creation of a new reserve. It is hoped that these admirable efforts will result in the Restinga Antwren's recovery, and in doing so set a precedent for the vital protection of other species facing similar threats.

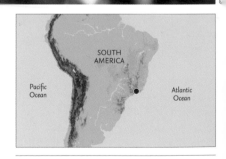

A species is *Critically Endangered* when it faces an extremely high risk of extinction in the wild, based on measurements of population size and/or geographic range and their trends in the past, present and/or future.

Djibouti Francolin

Francolinus ochropectus

NOT EVALUATED	DATA DEFICIENT	LEAST CONCERN	NEAR THREATENED	VULNERABLE	ENDANGERED	CRITICALLY ENDANGERED	EXTINCT IN THE WILD	EXTINCT
NE	DD	LC	NT	VU	EN	CR	EW	EX

The **Djibouti Francolin**, *Francolinus ochropectus*, is listed as 'Critically Endangered' on the IUCN Red List of Threatened Species™. It prefers the cover of dense vegetation found in the woodlands of Forêt du Day and the Mabla Mountains, Djibouti.

Djibouti Francolins are struggling to survive in Forêt du Day where the trees are dying if not already dead. This may be a result of high levels of grazing (the forest suddenly had to contend with year-round grazing once the previously nomadic Afars settled in the area) possibly exacerbated by acid rain and climate change. At Mabla, the remaining stands of woodland have been heavily exploited for firewood and grazing.

Work is now underway to protect Forêt du Day. To raise local awareness, brochures have been produced and distributed widely within schools, tourist centres and government departments. Plans for a community-based juniper forest restoration project are also underway.

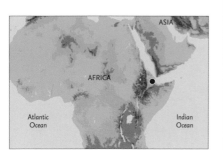

© Sam Cartwright

Mindoro Bleeding-heart

Gallicolumba platenae

NOT EVALUATED	DATA DEFICIENT	LEAST CONCERN	NEAR THREATENED	VULNERABLE	ENDANGERED	CRITICALLY ENDANGERED	EXTINCT IN THE WILD	EXTINCT
NE	DD	LC	NT	VU	EN	CR	EW	EX

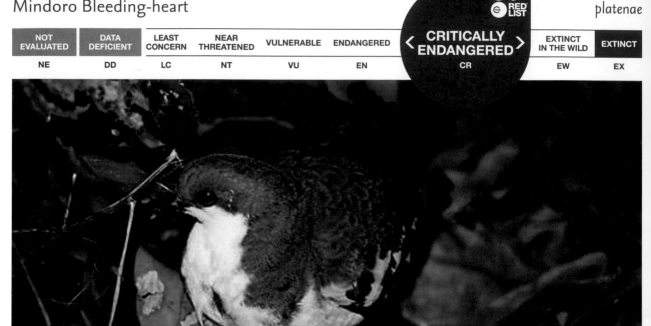

The **Mindoro Bleeding-heart**, *Gallicolumba platenae*, is listed as 'Critically Endangered' on the IUCN Red List of Threatened Species™. It is a shy, ground-dwelling pigeon that feeds on seeds, fallen fruit and worms found on the forest floor, and is native only to the island of Mindoro, in the Philippines.

Lowland forest destruction has eradicated almost all of the Mindoro Bleeding-heart's habitat. Forests at Siburan and Mount Iglit-Baco National Park are being threatened by encroaching shifting cultivation, slash-and-burn agriculture and occasional selective logging, whilst the collection of Rattan (climbing) palms disturbs the forest undergrowth. In forests at Puerto Galera, the Mindoro Bleeding-heart faces destruction from dynamite blasting for marble extraction. Because it forages on the forest floor, this species is also highly susceptible to being trapped in snares.

The continuation of studies that assess the Mindoro Bleeding-heart's requirements for breeding and foraging will help conserve this species and increase its survival rates. Other conservation measures needed include the elimination of illegal logging activities, regulation of hunting and forest-product extraction, and support for Mount Iglit-Baco National Park.

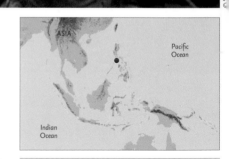

A species is **Critically Endangered** when it faces an extremely high risk of extinction in the wild, based on measurements of population size and/or geographic range and their trends in the past, present and/or future.

La Gomera Giant Lizard

Gallotia bravoana

NOT EVALUATED	DATA DEFICIENT	LEAST CONCERN	NEAR THREATENED	VULNERABLE	ENDANGERED	CRITICALLY ENDANGERED	EXTINCT IN THE WILD	EXTINCT
NE	DD	LC	NT	VU	EN	CR	EW	EX

© Dr. José A. Mateo

The **La Gomera Giant Lizard**, *Gallotia bravoana*, is listed as 'Critically Endangered' on the IUCN Red List of Threatened Species™. It is native to La Gomera Island (of the Canary Islands) and can grow to up to 1.2 metres in length.

This species was once widespread, living in a variety of places on La Gomera, but populations have declined through overgrazing, hunting, and predation by feral cats and rats. It was thought to be extinct, but in 1999 a small population was discovered to be living on a western part of the island. Although populations have been increasing since 2001 due to conservation efforts, they are severely fragmented in their distribution and only 90 individuals now remain in the wild.

There is now international legislation protecting La Gomera Giant Lizards, and a captive breeding programme has been established on the island. There is also a need to control cat populations on La Gomera and to implement education programmes for local people.

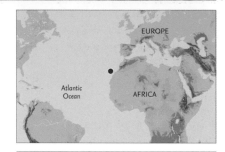

A species is Critically Endangered when it faces an extremely high risk of extinction in the wild, based on measurements of population size and/or geographic range and their trends in the past, present and/or future.

NOT EVALUATED	DATA DEFICIENT	LEAST CONCERN	NEAR THREATENED	‹ VULNERABLE ›	ENDANGERED	CRITICALLY ENDANGERED	EXTINCT IN THE WILD	EXTINCT
NE	DD	LC	NT	VU	EN	CR	EW	EX

Geonoma epetiolata is not currently listed on the IUCN Red List of Threatened Species™, however it has provisionally been assessed as 'Vulnerable'. This small, understorey palm is endemic to Costa Rica and Panama, where it is found within lowland and premontane rainforest. The young leaves are spectacularly coloured with yellows, greens, reds and purples, earning this species the name 'Stained Glass Palm'.

Habitat loss and collection for the ornamental plant trade are the greatest threats to this slender palm. The condition of the population in the type locality, near Santa Fe de Veraguas, Panama, is unknown.

In Panama, a population of Geonoma epetiolata is protected in Omar Torrijos National Park, while in Costa Rica the species is protected in a few privately owned nature reserves and within Braulio Carrillo National Park. However, despite legal protection, the illegal collection of this species for the horticultural trade still occurs within these areas.

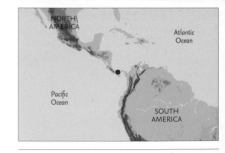

A species is Vulnerable when it faces a high risk of extinction in the wild, based on measurements of population size and/or geographic range and their trends in the past, present and/or future.

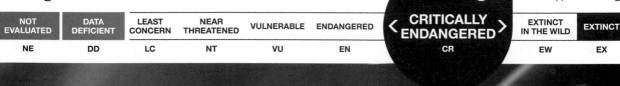

NOT EVALUATED	DATA DEFICIENT	LEAST CONCERN	NEAR THREATENED	VULNERABLE	ENDANGERED	CRITICALLY ENDANGERED	EXTINCT IN THE WILD	EXTINCT
NE	DD	LC	NT	VU	EN	CR	EW	EX

Geothlypis beldingi

© JAVIER LASCURAIN

Belding's Yellowthroat, Geothlypis beldingi, is listed as 'Critically Endangered' on the IUCN Red List of Threatened Species™. This attractively-marked warbler is restricted to a small and fragmented area of Baja California, Mexico, where it may number no more than a few thousand individuals.

The freshwater habitats occupied by Belding's Yellowthroat are under threat from human activities, including burning, reed-cutting, drainage for agriculture and cattle ranching, fire and overextraction of water. Recent surveys have found the species to occur at more sites than previously thought, but its tiny range and small, fragmented population make it vulnerable to any extreme events, such as hurricanes.

Part of the species' habitat falls within the Estero de San José del Cabo Ecological Reserve, a designated Ramsar site, where various conservation efforts are underway. Recommended conservation measures for this species include further surveys, genetic studies, greater habitat protection, and the creation of new areas of marshland. Reintroductions have also been suggested, while tourism may be a means by which income can be generated for the protection of key areas of habitat.

A species is *Critically Endangered* when it faces an extremely high risk of extinction in the wild, based on measurements of population size and/or geographic range and their trends in the past, present and/or future.

NOT EVALUATED	DATA DEFICIENT	LEAST CONCERN	NEAR THREATENED	VULNERABLE	ENDANGERED	‹ CRITICALLY ENDANGERED ›	EXTINCT IN THE WILD	EXTINCT
NE	DD	LC	NT	VU	EN	CR	EW	EX

Gladiolus balensis Goldblatt is listed as 'Critically Endangered' on the IUCN Red List of Threatened Species™. This plant is endemic to Ethiopia and thus far has only been collected in two locations. It grows in basaltic rocky areas which have now become largely cultivated or, in some areas, left fallow for grazing. This plant grows to about 55 cm in height, bearing a corm of about 20 mm in diameter, and has a very narrow altitudinal range of about 1,700–1,900 metres above sea level.

Gladiolus balensis Goldblatt is faced with a serious threat from agricultural expansion. In the two areas in which this species has been collected, conversion of woodland and wooded grassland into wheat and teff (Eragrostis tef) cultivation is occurring. Agricultural expansion and overgrazing by livestock affect both the extent of occurrence and area of occupancy of the species.

The two known populations of Gladiolus balensis Goldblatt are not in protected areas and there are no specific conservation efforts in place for this species.

A species is Critically Endangered when it faces an extremely high risk of extinction in the wild, based on measurements of population size and/or geographic range and their trends in the past, present and/or future.

Liberian Tree Hole Crab

Globonautes macropus

NOT EVALUATED	DATA DEFICIENT	LEAST CONCERN	NEAR THREATENED	VULNERABLE	❮ ENDANGERED ❯	CRITICALLY ENDANGERED	EXTINCT IN THE WILD	EXTINCT
NE	DD	LC	NT	VU	EN	CR	EW	EX

© Rüdiger Sachs and Neil Cumberlidge

The **Liberian Tree Hole Crab**, *Globonautes macropus*, is listed as 'Endangered' on the IUCN Red List of Threatened Species™. This unusual tree-climbing freshwater crab is endemic to the Upper Guinea forests of West Africa. It lives in water-filled tree holes in closed-canopy primary rainforest in Liberia and Guinea.

These air-breathing land crabs live well away from permanent water sources, forage at night on the forest floor, and climb tree trunks to hide deep inside holes where rainwater collects. Unfortunately, the rainforest habitats on which these crabs depend for their survival are under serious threat, and deforestation is increasing in all parts of the Upper Guinea forest.

These trends are driven by rising human populations and political unrest, resulting in pressures on the forest from the expansion of agriculture, firewood collection, logging and mining. The Liberian Tree Hole Crab is not found in a protected area, and this casts doubt on the long-term survival of this endangered and ecologically unusual species.

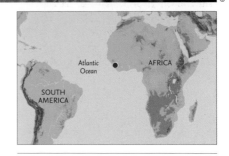

A species is Endangered when it faces a very high risk of extinction in the wild, based on measurements of population size and/or geographic range and their trends in the past, present and/or future.

Bog Turtle

NOT EVALUATED	DATA DEFICIENT	LEAST CONCERN	NEAR THREATENED	VULNERABLE	ENDANGERED	CRITICALLY ENDANGERED	EXTINCT IN THE WILD	EXTINCT
NE	DD	LC	NT	VU	EN	CR	EW	EX

The **Bog Turtle**, Glyptemys muhlenbergii, is listed as 'Critically Endangered' on the IUCN Red List of Threatened Species™. With a maximum shell length of 11 cm, this is one of the smallest turtle species in the world. When young, the shell of the bog turtle has a rough texture, but this is worn almost entirely smooth with age, the result of much burrowing. It occurs in small, scattered and fragmented areas of spring-fed wetlands in the northeastern United States and southern Appalachian foothills.

Severe habitat loss due to draining and conversion to agricultural lands has been, and continues to be, a primary reason for the decline of this species, with the added impacts of agricultural pollutants and changing climate patterns affecting water quality and availability. With its rich chocolate-brown shell, bright orange head blotches, and small size, the Bog Turtle has also been in great demand for the pet trade.

Recent reports of disease outbreaks give additional cause for grave concern for the future of this rare species. Strict protective legislation from the US Endangered Species Act and CITES Appendix I listing prohibiting possession and trade have provided some protection for the Bog Turtle.

A species is Critically Endangered when it faces an extremely high risk of extinction in the wild, based on measurements of population size and/or geographic range and their trends in the past, present and/or future.

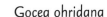

NOT EVALUATED	DATA DEFICIENT	LEAST CONCERN	NEAR THREATENED	VULNERABLE	ENDANGERED	CRITICALLY ENDANGERED	EXTINCT IN THE WILD	EXTINCT
NE	DD	LC	NT	VU	EN	CR	EW	EX

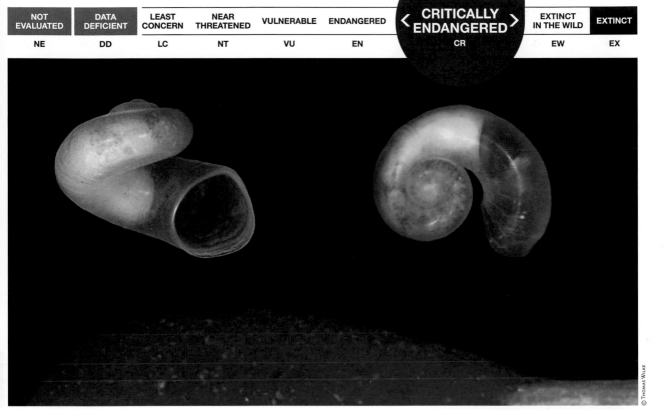

© Thomas Wilke

Gocea ohridana is listed as 'Critically Endangered' on the IUCN Red List of Threatened Species™. This species is endemic to the Macedonian part of ancient Lake Ohrid in the Balkans. It lives exclusively in the rocky southeastern shore habitats of the lake and is found in small numbers. Today, Gocea ohridana is restricted to a few square metres in the immediate surroundings of underwater springs found on the bottom of the lake. Within this setting the species is specialised, living in little holes in underwater rocks.

Deforestation has led to increased sediment loads entering the lake, blanketing the rocks and their holes, which affects this tiny species. Other threats come from pollution from settlements along Lake Ohrid, which contaminate the water. The underwater springs are fed by underground connections from Lake Prespa. Agriculture around this lake has increased the nutrient load in the waters which feed through to pollute Lake Ohrid and the springs.

Currently, there are no particular conservation measures in place to protect this species, although Lake Ohrid is a World Heritage Site.

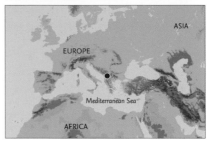

A species is Critically Endangered when it faces an extremely high risk of extinction in the wild, based on measurements of population size and/or geographic range and their trends in the past, present and/or future.

NOT EVALUATED	DATA DEFICIENT	LEAST CONCERN	NEAR THREATENED	VULNERABLE	ENDANGERED	CRITICALLY ENDANGERED	EXTINCT IN THE WILD	EXTINCT
NE	DD	LC	NT	VU	EN	CR	EW	EX

The **Field Cricket**, *Gryllus campestris*, is currently not officially listed on the IUCN Red List of Threatened Species™, however the conservation status has potentially been assessed as 'Least Concern'. It occurs from western Africa and northern Europe to the Caucasus region.

In the northern part of its range, including many parts of Europe, the Field Cricket is declining due to habitat loss, degradation and fragmentation. In this region, it is dependent on dry, nutrient-poor habitats with short grass or heather vegetation. Such habitat types have become rare and fragmented due to agricultural intensification and increasing eutrophication.

Traditional sheep herding ceased at the end of the 19th century, and former heathland habitats have been replaced by coniferous forest or farmland.

The Field Cricket is a good indicator of species-rich habitats (its conspicuous song is familiar to most people on warm summer nights). It has been successfully reintroduced in the UK and parts of Germany. However, the most important conservation priority is found in traditional land use, including sheep herding and heathland management.

A species is Least Concern when it does not qualify for Critically Endangered, Endangered, Vulnerable or Near Threatened. Widespread and abundant species are included in this category.

White-rumped Vulture

Gyps bengalensis

NOT EVALUATED	DATA DEFICIENT	LEAST CONCERN	NEAR THREATENED	VULNERABLE	ENDANGERED	CRITICALLY ENDANGERED	EXTINCT IN THE WILD	EXTINCT
NE	DD	LC	NT	VU	EN	CR	EW	EX

A species is **Critically Endangered** when it faces an **extremely high risk** of extinction in the wild, based on measurements of population size and/or geographic range and their trends in the past, present and/or future.

© ROY DE HAAS/WWW.AGAMI.NL

The **White-rumped Vulture**, Gyps bengalensis, is listed as 'Critically Endangered' on the IUCN Red List of Threatened Species™. Large numbers used to be seen soaring above villages, towns and cities throughout southern Asia, and it was thought to be the most abundant large bird of prey in the world – probably numbering several million individuals. Now, less than 10,000 exist in the wild.

The abrupt population decline was thought to be due to a fatal virus, but testing revealed it was attributable to the effects of feeding on carcasses treated with the veterinary drug diclofenac. Local people are affected by the decline in the number of White-rumped Vultures, as rotting carcasses remain untouched, creating potential health hazards or even attracting rabid feral dogs.

Captive breeding programmes have been set up alongside initiatives to help raise public awareness of this Critically Endangered vulture. A substitute drug has been found to replace diclofenac, which has been banned on the Indian subcontinent, but stocks are still available and its use has not yet been completely eliminated.

NOT EVALUATED	DATA DEFICIENT	LEAST CONCERN	NEAR THREATENED	VULNERABLE	ENDANGERED	CRITICALLY ENDANGERED	EXTINCT IN THE WILD	EXTINCT
NE	DD	LC	NT	VU	EN	CR	EW	EX

The **Indian Vulture**, Gyps indicus, is listed as 'Critically Endangered' on the IUCN Red List of Threatened Species™. This robust and scruffy scavenger used to be found in villages, towns and cities near cultivated and wooded areas in southeastern Pakistan and India.

Like other vulture species, the Indian Vulture has suffered serious declines since the late 1990s, losing as much as 99 percent of its population. Vultures were discovered to be suffering from kidney failure following the consumption of cattle that had previously been treated with the anti-inflammatory drug diclofenac. As a result of reduced vulture numbers, rotting animal carcasses now remain untouched, causing health hazards, as well as encouraging feral dog populations which may carry rabies.

Strong government commitment has been given to the prevention of any use of this drug and for its complete removal from the environment, but supplies are still being used, sometimes in ignorance of their effects. The Indian Vulture will also benefit from a new captive breeding programme, raised awareness and further monitoring.

A species is Critically Endangered when it faces an extremely high risk of extinction in the wild, based on measurements of population size and/or geographic range and their trends in the past, present and/or future.

Slender-billed Vulture

NOT EVALUATED	DATA DEFICIENT	LEAST CONCERN	NEAR THREATENED	VULNERABLE	ENDANGERED	CRITICALLY ENDANGERED	EXTINCT IN THE WILD	EXTINCT
NE	DD	LC	NT	VU	EN	CR	EW	EX

A species is Critically Endangered when it faces an extremely high risk of extinction in the wild, based on measurements of population size and/or geographic range and their trends in the past, present and/or future.

The **Slender-billed Vulture**, Gyps tenuirostris, is listed as 'Critically Endangered' on the IUCN Red List of Threatened Species™. Found in India, Bangladesh and parts of Southeast Asia, this once-common species has undergone a catastrophic decline in the last decade.

By the late 1990s, dead and dying vultures, of this and two other Gyps species, were being found across many areas of the subcontinent. It took five years of intensive research to determine that the cause of the decline was the anti-inflammatory drug diclofenac, which became available in India around 1994 as a veterinary drug for use on sick livestock.

Various laws banning the manufacture and import of diclofenac are now in place, and a suitable alternative has been found and is being promoted, but stocks of diclofenac still exist. Slender-billed Vultures are also being bred in captivity, and vulture 'restaurants' are being used as ecotourism attractions, to raise awareness and provide supplementary feeding. The recovery of this species will also benefit humans, as vultures clear up rotting carcasses which may otherwise present a health hazard.

Social Wrasse

NOT EVALUATED	DATA DEFICIENT	LEAST CONCERN	NEAR THREATENED	VULNERABLE	ENDANGERED	CRITICALLY ENDANGERED	EXTINCT IN THE WILD	EXTINCT
NE	DD	LC	NT	VU	EN	CR	EW	EX

The **Social Wrasse**, Halichoeres socialis is listed as 'Critically Endangered' on the IUCN Red List of Threatened Species™. The Social Wrasse is only found in the Pelican Cays of Belize, with a total range of less than 10 km².

This species is heavily reliant on mangrove wetlands and coral reefs for spawning as well as feeding. In the past 10 years the Belize tourism industry has skyrocketed, and people from all over the world are spending holidays there to enjoy the beautiful beaches and great diving. However, this growing industry has also led to the destruction of many mangrove and coral reef habitats, for development, off mainland Belize and the Pelican Cays. This rapid development and habitat destruction could further limit the already restricted range of the Social Wrasse, and possibly lead to its extinction.

Although the Pelican Cays are classified as a World Heritage Site, there are no measures to prevent further habitat destruction there. The Social Wrasse must be closely monitored and studied in order to develop protective protocols that can save this species and its habitat.

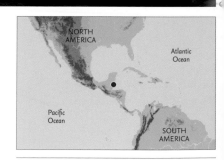

A species is *Critically Endangered* when it faces an extremely high risk of extinction in the wild, based on measurements of population size and/or geographic range and their trends in the past, present and/or future.

Black Abalone

NOT EVALUATED	DATA DEFICIENT	LEAST CONCERN	NEAR THREATENED	VULNERABLE	ENDANGERED	CRITICALLY ENDANGERED	EXTINCT IN THE WILD	EXTINCT
NE	DD	LC	NT	VU	EN	CR	EW	EX

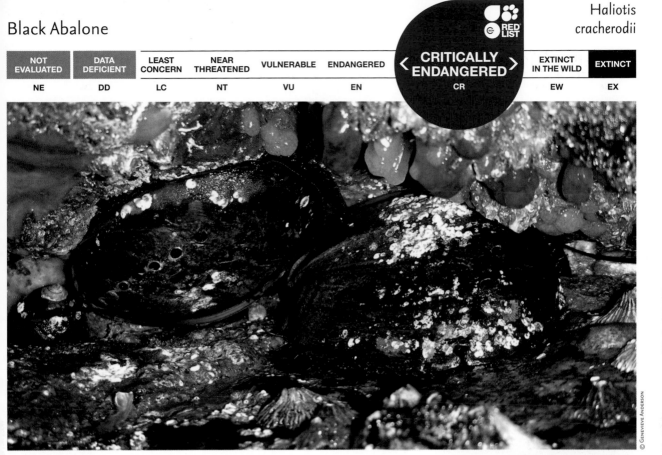

© Genevieve Anderson

The **Black Abalone**, Haliotis cracherodii, is listed as 'Critically Endangered' on the IUCN Red List of Threatened Species™. This large marine mollusc is found in rocky intertidal and subtidal habitats on the coasts of Baja California, Mexico and California, USA.

Over the last few decades, the Black Abalone has suffered a serious decline, primarily due to a disease known as withering syndrome, which has spread through a large proportion of the population. The disease causes wasting of the foot muscle, preventing the abalone from properly adhering to the substrate, causing it to become discoloured, lose weight and die. Other threats include commercial and recreational fishing, coastal development and pollution.

All abalone fisheries in California are managed with seasonal closures and restrictions on catch size, while in Mexico there is a total allowable catch limit. In January 2009, the Black Abalone was finally listed under the US Endangered Species Act (ESA). Under the ESA, this species should benefit from the compulsory development of a recovery plan, protection and restoration of critical habitat, scientific research and public education.

A species is Critically Endangered when it faces an extremely high risk of extinction in the wild, based on measurements of population size and/or geographic range and their trends in the past, present and/or future.

Sea Clover

NOT EVALUATED	DATA DEFICIENT	LEAST CONCERN	NEAR THREATENED	⟨ VULNERABLE ⟩	ENDANGERED	CRITICALLY ENDANGERED	EXTINCT IN THE WILD	EXTINCT
NE	DD	LC	NT	VU	EN	CR	EW	EX

Sea Clover, Halophila baillonii, is listed as 'Vulnerable' on the IUCN Red List of Threatened Species™. It is a species of seagrass, which are vascular flowering plants that live submerged in the world's oceans. Sea Clover has a severely fragmented distribution in the Caribbean Sea, with recent disappearances from Brazil and the Pacific coast of Costa Rica.

Sea Clover is an important food source for the West Indian Manatee, a large sea mammal that is itself endangered, as well as the endangered Green Turtle. It is very sensitive to poor water quality and is threatened by coastal development. Runoff from aquaculture, agriculture and industry are major threats. In Belize, the largest known meadow of Sea Clover is rapidly declining as a result of unregulated tourist development and eutrophication along Placencia Lagoon. In Costa Rica, the only known bed of this seagrass was destroyed by a severe Pacific storm.

There are currently no specific conservation measures for Sea Clover, although it sometimes occurs within Marine Protected Areas. Both protection and monitoring of this ecologically important seagrass species are imperative.

A species is **Vulnerable** when it faces a high risk of extinction in the wild, based on measurements of population size and/or geographic range and their trends in the past, present and/or future.

Ocean Turf Grass

Halophila beccarii

NOT EVALUATED	DATA DEFICIENT	LEAST CONCERN	NEAR THREATENED	‹ VULNERABLE ›	ENDANGERED	CRITICALLY ENDANGERED	EXTINCT IN THE WILD	EXTINCT
NE	DD	LC	NT	VU	EN	CR	EW	EX

© Fred Short / www.SeagrassNet.org

Ocean Turf Grass, *Halophila beccarii*, is classified as 'Vulnerable' on the IUCN Red List of Threatened Species™. It lives submerged in the world's oceans, and its distribution is fragmented in southern China, Southeast Asia, India and Madagascar, amongst seaward mangroves and lagoons, in estuaries and on mudflats. Ocean Turf Grass is a food source for marine invertebrates and some fishes, and a habitat for juvenile horseshoe crabs.

Ocean Turf Grass is particularly susceptible to threats, as it has a restricted habitat type in the shallow intertidal zone where there is often human activity. The major threats to this seagrass are deforestation of mangroves and the creation of shrimp aquaculture ponds, which eliminate the areas where it thrives, along with coastal development, harbour and jetty construction, sedimentation, sand mining, and destructive fishing methods.

There are currently no specific conservation measures for this species, except when it occurs within Marine Protected Areas. Shoreline protection and reduction in shoreline hardening, along with preservation and restoration of mangroves, will be important to its long-term survival.

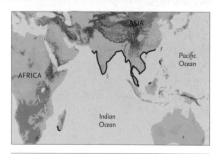

A species is *Vulnerable* when it faces a high risk of extinction in the wild, based on measurements of population size and/or geographic range and their trends in the past, present and/or future.

NOT EVALUATED	DATA DEFICIENT	LEAST CONCERN	NEAR THREATENED	VULNERABLE	ENDANGERED	CRITICALLY ENDANGERED	EXTINCT IN THE WILD	EXTINCT
NE	DD	LC	NT	VU	EN	CR	EW	EX

Halophila nipponica is listed as 'Near Threatened' on the IUCN Red List of Threatened Species™. This seagrass is found in a few temperate locations in Japan, and one site in southern South Korea. A small seagrass, just 5 cm tall, Halophila nipponica is often found growing on its own, or as an understorey to Zostera marina or other long-bladed species. The small, rounded leaves of this species provide sediment stabilization and removal of dissolved nutrients.

Halophila nipponica is declining in the industrial and urban areas of its range. Healthy seagrass beds indicate healthy coastal waters, but they are directly threatened by anthropogenic inputs. In Japan, this species is subject to threats from coastal development, land reclamation, water pollution and fisheries trawling. A single population was recently found in South Korea, occupying a large meadow near offshore islands in six metres of water. This recent appearance may be related to increasing seawater temperatures.

There are no known specific conservation measures for Halophila nipponica. Research is needed on this species' taxonomy, population trends and threats.

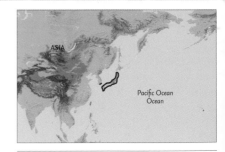

A species is **Near Threatened** when it does not qualify for Critically Endangered, Endangered or Vulnerable now, but is close to qualifying for or is likely to qualify for a threatened category in the near future.

Alaotran Gentle Lemur

NOT EVALUATED	DATA DEFICIENT	LEAST CONCERN	NEAR THREATENED	VULNERABLE	ENDANGERED	CRITICALLY ENDANGERED	EXTINCT IN THE WILD	EXTINCT
NE	DD	LC	NT	VU	EN	CR	EW	EX

The **Alaotran Gentle Lemur**, Hapalemur alaotrensis, is classified as 'Critically Endangered' on the IUCN Red List of Threatened Species™. Fewer than 3,000 individuals remain in the wild in two fragmented populations within 19,000 hectares of marsh in Lake Alaotra, the largest lake in Madagascar. The Alaotran Gentle Lemur is the only exclusively wetland-dwelling primate in the world.

The main threats to the species are loss of habitat to rice agriculture, uncontrolled marsh burning during the dry season, and poaching. Burning is largely carried out to access new areas for fishing, but as well as destroying habitat it causes direct mortality to native species and facilitates the spread of invasive plants.

The entire Lake Alaotra watershed was designated as a Ramsar site in 2004, and the lake and marshes were made a protected area in 2007. Conservation organizations have been working closely with local communities to encourage sustainable use of the wetland and to reduce poaching pressure on the lemur. Rigorous protection of the Lake Alaotra marshes and engagement with local communities must be maintained in order to save the Alaotran Gentle Lemur from extinction.

© James Morgan

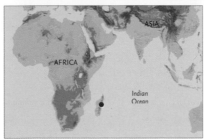

A species is **Critically Endangered** when it faces an extremely high risk of extinction in the wild, based on measurements of population size and/or geographic range and their trends in the past, present and/or future.

Sun Bear

NOT EVALUATED	DATA DEFICIENT	LEAST CONCERN	NEAR THREATENED	‹ VULNERABLE ›	ENDANGERED	CRITICALLY ENDANGERED	EXTINCT IN THE WILD	EXTINCT
NE	DD	LC	NT	VU	EN	CR	EW	EX

The **Sun Bear**, *Helarctos malayanus*, is listed as 'Vulnerable' on the IUCN Red List of Threatened Species™. It occurs in forest patches of mainland Southeast Asia, Sumatra and Borneo. The world's smallest bear, it is identified by an individually distinct patch of light fur, resembling the sun, on the chest.

The Sun Bear's range has been greatly fragmented and reduced through large-scale deforestation, due to legal and illegal logging and burning, and conversion of natural forests to oil palm and other commercial crops. It is also illegally exploited for its body parts, which are used in traditional medicines, and killed when ransacking human crops. Although the killing of Sun Bears is strictly prohibited, the laws are seldom enforced.

This species is relatively poorly studied, and little conservation action has been targeted directly at it. Although no population estimates are currently available, it has become clear from the rapid loss of habitat that Sun Bear numbers must be in significant decline. Protection of habitat and reduction of poaching are paramount to its conservation.

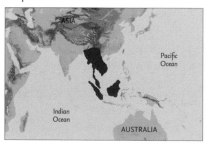

© G. Fredriksson

Table Mountain Ghost Frog

Heleophryne rosei

NOT EVALUATED	DATA DEFICIENT	LEAST CONCERN	NEAR THREATENED	VULNERABLE	ENDANGERED	CRITICALLY ENDANGERED	EXTINCT IN THE WILD	EXTINCT
NE	DD	LC	NT	VU	EN	CR	EW	EX

© VINCENT CARRUTHERS & LOUIS DU PREEZ

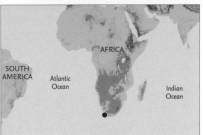

A species is **Critically Endangered** when it faces an extremely high risk of extinction in the wild, based on measurements of population size and/or geographic range and their trends in the past, present and/or future.

The **Table Mountain Ghost Frog**, Heleophryne rosei, is listed as 'Critically Endangered' on the IUCN Red List of Threatened Species™. This elusive species is restricted to swift-flowing streams on the slopes of Cape Town's iconic Table Mountain.

The Table Mountain Ghost Frog is subject to numerous threats, each with the potential to have a devastating impact given this species' narrow range. The spread of alien vegetation has resulted in the clogging of streams, while the construction of dams has reduced water flow, creating areas of stagnant water. Other potential threats include frequent fires, climate change, and ecotourism.

The whole of this species' range lies within the Table Mountain National Park, a part of the Cape Floristic World Heritage Site. One of the main conservation priorities is to ensure the preservation of swift-flowing perennial streams on Table Mountain.

Inagua Rockrose

NOT EVALUATED	DATA DEFICIENT	LEAST CONCERN	NEAR THREATENED	VULNERABLE	ENDANGERED	CRITICALLY ENDANGERED	EXTINCT IN THE WILD	EXTINCT
NE	DD	LC	NT	VU	EN	CR	EW	EX

A species is **Critically Endangered** when it faces an extremely high risk of extinction in the wild, based on measurements of population size and/or geographic range and their trends in the past, present and/or future.

The **Inagua Rockrose**, *Helianthemum inaguae*, is classified as 'Critically Endangered' on the Red List of Spanish Vascular Flora. It is exclusive to the Canary Islands (Gran Canaria), where it has only been found in a small area located in the western sector of the island. In July 2007, a large forest fire seriously affected this area, destroying the majority of plants there.

In addition to forest fires, the impact of herbivores appears to be a major threat to this species. A certain amount of habitat fragmentation and low levels of recruitment, due to increasingly frequent periods of drought, have also been observed.

The species is protected regionally and nationally. It is listed as an Endangered Species in the National Catalogue of Endangered Species, as well as in the Canary Islands Catalogue of Endangered Species, and a recovery plan needs to be developed. The unrestrained livestock in the area where the species lives need be controlled, and specific activities to reinforce the natural population, as well as translocations to other optimal habitats, should be carried out.

Mushroom Coral

NOT EVALUATED	DATA DEFICIENT	LEAST CONCERN	NEAR THREATENED	⟨ VULNERABLE ⟩	ENDANGERED	CRITICALLY ENDANGERED	EXTINCT IN THE WILD	EXTINCT
NE	DD	LC	NT	VU	EN	CR	EW	EX

RED LIST

© Mark Spencer / Auscape International

The **Mushroom Coral**, *Heliofungia actiniformis*, is listed as 'Vulnerable' on the IUCN Red List of Threatened Species™. It is closely related to only one other species and is found in the Indo-Pacific region, in Southeast Asia and Australia. Shrimp and fish species live among the tentacles of Mushroom Coral.

The Mushroom Coral is popular in the aquarium trade and is threatened by unsustainable levels of harvesting. Unmonitored tourism, destructive fishing, coastal development, sedimentation and sewage from growing human populations also threaten this species. Additionally, coral bleaching and ocean acidification, caused by climate change, are major global threats to corals.

The Mushroom Coral is protected by CITES, which regulates and bans the export of all corals. Currently, Indonesia is the only country legally allowed to export this species. Marine Protected Areas and coral conservation programmes also occur within the range of Mushroom Coral.

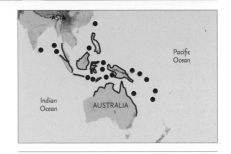

A species is Vulnerable when it faces a high risk of extinction in the wild, based on measurements of population size and/or geographic range and their trends in the past, present and/or future.

Blue Coral

Heliopora coerulea

Blue Coral, *Heliopora coerulea*, is listed as 'Vulnerable' on the IUCN Red List of Threatened Species™. Named for its permanently blue skeleton, which makes it popular in the curio, jewellery and aquarium trades, Blue Coral is very widespread in the Indo-Pacific region, but only occurs in limited areas on extremely shallow reefs.

In addition to the threat from over-collection, Blue Coral is particularly susceptible to the rise in sea temperature associated with global climate change. This leads to coral bleaching, where the symbiotic algae are expelled, leaving the coral weak and vulnerable to an increasing variety of harmful diseases. More localised threats include pollution, destructive fishing practices, invasive species, human development, and other activities.

In addition to being listed on Appendix II of CITES, which makes it an offence to trade Blue Coral internationally without a permit, this species falls within several Marine Protected Areas. To conserve Blue Coral, studies into various aspects of its taxonomy, biology and ecology, including assessing threats and potential recovery techniques, have been recommended.

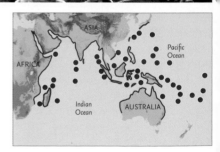

*A species is **Vulnerable** when it faces a high risk of extinction in the wild, based on measurements of population size and/or geographic range and their trends in the past, present and/or future.*

Heritiera globosa

NOT EVALUATED	DATA DEFICIENT	LEAST CONCERN	NEAR THREATENED	VULNERABLE	‹ ENDANGERED ›	CRITICALLY ENDANGERED	EXTINCT IN THE WILD	EXTINCT
NE	DD	LC	NT	VU	EN	CR	EW	EX

© J. E. Ong

The **Globosa Mangrove**, *Heritiera globosa*, is listed as 'Endangered' on the IUCN Red List of Threatened Species™. This species of mangrove is one of Indonesia's many endemic species and is only found in the West Kalimantan province, on the island of Borneo. It is a very rare species of mangrove that has a very patchy distribution within a specific salinity range and prefers the more freshwater-dominated areas of upstream riverine habitats. It is a slow-growing species, but can reach large sizes of up to 25 metres.

Like many other species of mangrove, the Globosa is being depleted at an alarming rate. The Globosa Mangroves are suffering mostly due to habitat loss and logging. Many mangrove swamps, especially areas with higher concentrations of freshwater, are being targeted and destroyed, to be replaced with commercial timber and oil palm farms.

Currently this mangrove is not a protected species and is not found in any protected areas. It is not expected to survive the next several decades if its destruction is not curbed.

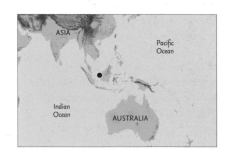

Liben Lark

NOT EVALUATED	DATA DEFICIENT	LEAST CONCERN	NEAR THREATENED	VULNERABLE	ENDANGERED	CRITICALLY ENDANGERED	EXTINCT IN THE WILD	EXTINCT
NE	DD	LC	NT	VU	EN	CR	EW	EX

The **Liben Lark**, *Heteromirafra sidamoensis*, is listed as 'Critically Endangered' on the IUCN Red List of Threatened Species™. Found only in one small area of open grassland in southern Ethiopia, this little-known bird has a tiny population, and could well become extinct within just a few years.

Its occurrence in a single small, unprotected area, the Liben Plain just outside Negele, makes the Liben Lark highly vulnerable to any threats. Currently, the greatest threat to the species is habitat loss, with an increasing number of livestock leading to problems with overgrazing and trampling, along with conversion of grassland to agriculture. Excessive grazing and fire suppression have also led to the rapid encroachment of shrubs.

Fieldwork carried out since 2007 to investigate the Liben Lark's status and requirements has revealed the bird's plight, and there is now a growing programme of conservation work led by the Ethiopian Wildlife and Natural History Society involving the local communities and focusing on reduction of grazing in some areas, clearance of invading scrub, and greater all-round awareness.

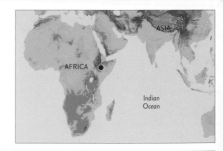

A species is Critically Endangered when it faces an extremely high risk of extinction in the wild, based on measurements of population size and/or geographic range and their trends in the past, present and/or future.

Giant Freshwater Stingray

NOT EVALUATED	DATA DEFICIENT	LEAST CONCERN	NEAR THREATENED	⟨ VULNERABLE ⟩	ENDANGERED	CRITICALLY ENDANGERED	EXTINCT IN THE WILD	EXTINCT
NE	DD	LC	NT	VU	EN	CR	EW	EX

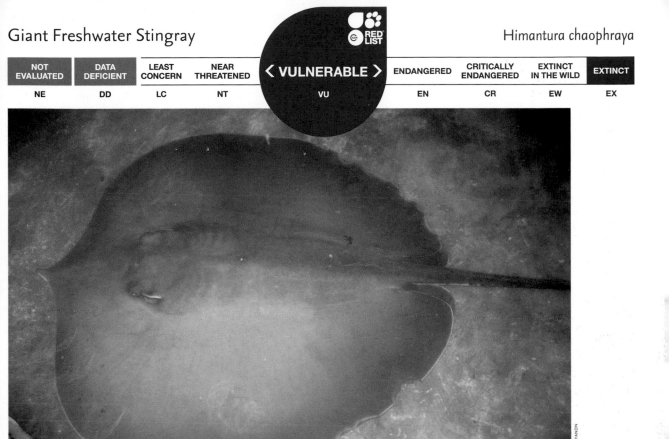

© Chavalit Vidthayanon

The **Giant Freshwater Stingray**, *Himantura chaophraya*, is listed as 'Vulnerable' on the IUCN Red List of Threatened Species™. One of the largest freshwater fish in the world, this impressive species occurs in several rivers in Southeast Asia and northern Australia.

The subpopulation of Giant Freshwater Stingrays in Thailand is listed as 'Critically Endangered', and is under threat from both direct fishing and accidental capture. The alteration and degradation of river habitat is a further threat, and results from a range of factors including deforestation, dam-building and pollution. Dams further threaten the species by isolating parts of the population, preventing interbreeding. While the species appears to be more secure in Australia, polluted silt from uranium mines is a potential cause for concern.

The dramatic decline of Giant Freshwater Stingrays in Thailand led to the establishment of an experimental captive propagation programme, aiming to stabilize populations while attempts are made to address habitat degradation. A national recovery team is planned in Australia, and further research into the species is urgently needed in other parts of its range.

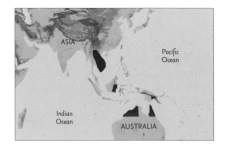

NOT EVALUATED	DATA DEFICIENT	LEAST CONCERN	NEAR THREATENED	VULNERABLE	‹ ENDANGERED ›	CRITICALLY ENDANGERED	EXTINCT IN THE WILD	EXTINCT
NE	DD	LC	NT	VU	EN	CR	EW	EX

© Jo Anne Smith and Werner Flueck

The **Huemul**, *Hippocamelus bisulcus*, is listed as 'Endangered' on the IUCN Red List of Threatened Species™. Once widespread across the southern Andes and Patagonia, this deer now has a much more restricted range within southern Argentina and Chile.

The Huemul's decline has been attributed to a range of factors, including overhunting, overstocking of its habitat with domestic livestock, land conversion to agriculture, and, more locally, threats such as construction, logging, poaching and disease. Suitable winter habitat has been almost entirely eliminated, and illegal hunting continues in many areas. The remaining subpopulations are small and fragmented, and the species may now total fewer than 1,500 individuals.

Although now fully protected in Chile and Argentina, most of the Huemul population occurs outside of protected areas, and laws are often poorly enforced even within reserves. Urgent conservation measures include raising awareness of its plight, promoting the creation of private protected areas, removing livestock in national parks, and captive breeding.

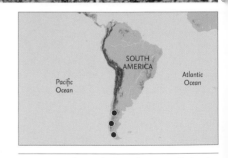

A species is **Endangered** when it faces a very high risk of extinction in the wild, based on measurements of population size and/or geographic range and their trends in the past, present and/or future.

NOT EVALUATED	DATA DEFICIENT	LEAST CONCERN	NEAR THREATENED	< VULNERABLE >	ENDANGERED	CRITICALLY ENDANGERED	EXTINCT IN THE WILD	EXTINCT
NE	DD	LC	NT	VU	EN	CR	EW	EX

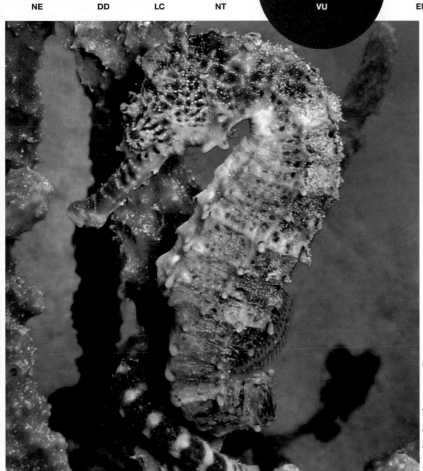

© ANTIDIO ROSSI, GUYLIAN SEAHORSES OF THE WORLD 2005

The **Tiger Tail Seahorse**, Hippocampus comes, is classified as 'Vulnerable' on the IUCN Red List of Threatened Species™. This elegant fish is known from the central and southern regions of the Philippines, and westward through Vietnam, Singapore, Malaysia and northern Indonesia to the Andaman Islands (India).

The Tiger Tail Seahorse is fished heavily for use in the aquarium trade and as 'traditional medicine'. Individuals are extracted from their habitats of coral reefs, seagrass beds and mangroves, through the use of nets and more harmful methods such as dynamite and cyanide poisoning. It is estimated that populations have diminished by almost 50%, and in some areas in the Philippines by as much as 85%. Their slow movement and limited larval dispersal make this species very vulnerable to localized exterminations.

The Tiger Tail Seahorse is listed on CITES Appendix II, which makes it an offense to trade this species internationally without a permit. However, this has done little to deter collection in its native range. Ongoing research and local protective initiatives from foundations such as Project Seahorse© are needed in order to prevent the loss of this graceful species.

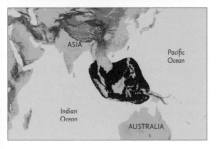

A species is Vulnerable when it faces a high risk of extinction in the wild, based on measurements of population size and/or geographic range and their trends in the past, present and/or future.

Lined Seahorse

Hippocampus erectus

NOT EVALUATED	DATA DEFICIENT	LEAST CONCERN	NEAR THREATENED	‹ VULNERABLE ›	ENDANGERED	CRITICALLY ENDANGERED	EXTINCT IN THE WILD	EXTINCT
NE	DD	LC	NT	VU	EN	CR	EW	EX

The **Lined Seahorse**, *Hippocampus erectus*, is listed as 'Vulnerable' on the IUCN Red List of Threatened Species™. It is the most commonly recognised species of seahorse, and can be found from Nova Scotia, Canada, along the east coast of North America and down to Brazil, as well as throughout the Gulf of Mexico and the Caribbean Sea.

Over the past several decades, the Lined Seahorse has come under extreme pressure from harvesting and loss of habitat. Every year, thousands of individuals are caught and sold in the aquarium trade and are increasingly being exported to China as traditional medicine. This species is also threatened by habitat destruction and by-catch in shrimp trawlers. As a result, it is estimated that the Lined Seahorse population has declined by at least 30% in the past 10 years.

Hippocampus erectus and other species in the genus Hippocampus have been listed under Appendix II of CITES since 2002, and there is strong monitoring of their harvest in the United States. Unfortunately, however, there is still widespread destruction and harvesting of this species.

A species is **Vulnerable** when it faces a high risk of extinction in the wild, based on measurements of population size and/or geographic range and their trends in the past, present and/or future.

Itatiaia Highland Frog

Holoaden bradei

NOT EVALUATED	DATA DEFICIENT	LEAST CONCERN	NEAR THREATENED	VULNERABLE	ENDANGERED	CRITICALLY ENDANGERED	EXTINCT IN THE WILD	EXTINCT
NE	DD	LC	NT	VU	EN	CR	EW	EX

© IVAN SAZIMA

The **Itatiaia Highland Frog**, *Holoaden bradei*, is listed as 'Critically Endangered' on the IUCN Red List of Threatened Species™. Restricted to just one small area of the Itatiaia Mountains in south-eastern Brazil, the species has not been recorded alive for about 30 years, and may already have become extinct.

This small frog has not been recorded outside of a single small area of under ten square kilometres, and may have declined as a result of a reduction in the quality of its habitat.

Tourism and fire have degraded parts of its open grassland and sparse forest habitat, but the reasons for its disappearance are unknown.

The range of the Itatiaia Highland Frog occurs within the Parque Nacional do Itatiaia, which should hopefully afford the species some protection. However, surveys are needed to determine whether this rare and little-known amphibian is still extant. If it is found, urgent conservation measures, including the continued preservation of its habitat, will be needed to save the remaining population.

A species is Critically Endangered when it faces an extremely high risk of extinction in the wild, based on measurements of population size and/or geographic range and their trends in the past, present and/or future.

Bengal Florican

NOT EVALUATED	DATA DEFICIENT	LEAST CONCERN	NEAR THREATENED	VULNERABLE	ENDANGERED	CRITICALLY ENDANGERED	EXTINCT IN THE WILD	EXTINCT
NE	DD	LC	NT	VU	EN	CR	EW	EX

The **Bengal Florican**, Houbaropsis bengalensis, is listed as 'Critically Endangered' on the IUCN Red List of Threatened Species™. It is found on the Indian subcontinent and in parts of Southeast Asia including Cambodia and southern Vietnam. During courtship the black-bodied male makes deep humming sounds while it leaps into the air, showing off its snow-white wings.

The Bengal Florican is primarily threatened by the loss and alteration of its grassland habitat. On the subcontinent it occurs almost exclusively in reserves, but the numbers are very small. In Cambodia, the grasslands around the Tonle Sap are rapidly being converted into rice monocultures. In Vietnam it is already believed to have disappeared. Excessive hunting for sport and food (particularly in Cambodia) has also played a part in the decline of this rare bird and remains a serious threat.

The national parks and wildlife reserves of India and Nepal have had some success in slowing the decline in Bengal Florican numbers, but its populations there are dwindling and more management of its grasslands is needed. Work in Cambodia is focused on establishing small grassland reserves where modern rice-growing is prohibited.

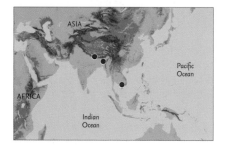

A species is **Critically Endangered** when it faces an extremely high risk of extinction in the wild, based on measurements of population size and/or geographic range and their trends in the past, present and/or future.

Beluga Sturgeon

Huso huso

NOT EVALUATED	DATA DEFICIENT	LEAST CONCERN	NEAR THREATENED	VULNERABLE	‹ ENDANGERED ›	CRITICALLY ENDANGERED	EXTINCT IN THE WILD	EXTINCT
NE	DD	LC	NT	VU	EN	CR	EW	EX

© TONY GILBERT

The **Beluga Sturgeon**, *Huso huso*, is listed as 'Endangered' on the IUCN Red List of Threatened Species™. Also known as the Giant Sturgeon, it is the largest sturgeon in the world, and the largest freshwater fish in Europe, where it occurs primarily in and around the Caspian, Black, and occasionally in the Adriatic and Mediterranean Seas.

Overfishing, habitat loss and pollution threaten the survival of the Beluga Sturgeon. Its eggs are highly prized as beluga caviar, and illegal fishing is reported to be common. Pollution, dams and silting can cause habitat destruction by damaging coastal waters and altering river courses. For example, the Volgograd Dam in Russia has effectively blocked access to almost all of the species' spawning grounds along the Volga River.

The majority of the Beluga Sturgeon population is now supported artificially, with hatcheries thought to be the sole reason the species still survives in the Caspian Sea. The United States is the biggest importer of beluga caviar, and has now listed the species as 'Threatened' under the Endangered Species Act, suspending imports.

A species is Endangered when it faces a very high risk of extinction in the wild, based on measurements of population size and/or geographic range and their trends in the past, present and/or future.

Brown Hyaena

Hyaena brunnea

			NEAR THREATENED						
NOT EVALUATED	DATA DEFICIENT	LEAST CONCERN		VULNERABLE	ENDANGERED	CRITICALLY ENDANGERED	EXTINCT IN THE WILD	EXTINCT
NE	DD	LC	NT	VU	EN	CR	EW	EX

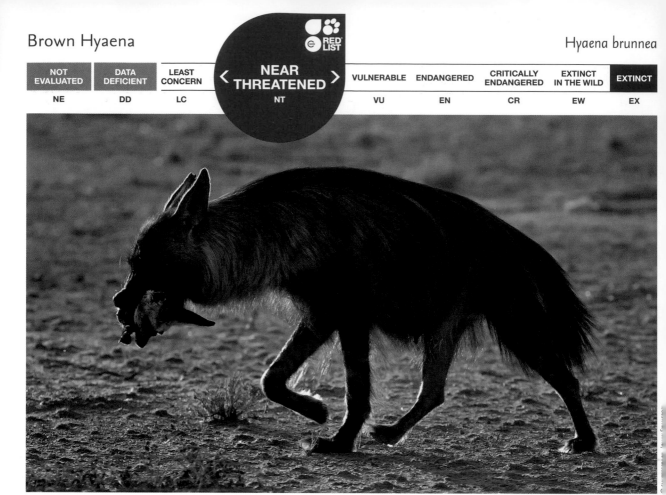

The **Brown Hyaena**, *Hyaena brunnea*, is listed as 'Near Threatened' on the IUCN Red List of Threatened Species™. Occurring in arid parts of southern Africa, the Brown Hyaena is distinguished from other hyaenas by its long, shaggy coat and pointed ears.

Unfortunately, the Brown Hyaena has been the victim of strong negative attitudes, wrongly being perceived as a threat to domestic livestock. It is often shot, poisoned, trapped, or hunted with dogs, as well as being inadvertently killed in non-selective predator control programmes.

As a result, this much-maligned species has suffered a decline in range and abundance.

The Brown Hyaena occurs in several large conservation areas throughout its range, although conflict with humans still occurs outside of these areas. Fortunately, education and awareness campaigns are starting to slowly change attitudes towards the Brown Hyaena, meaning that the future may be brighter for this charismatic but misunderstood species.

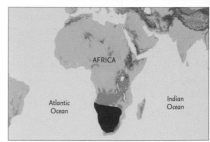

A species is Near Threatened when it does not qualify for Critically Endangered, Endangered or Vulnerable now, but is close to qualifying for or is likely to qualify for a threatened category in the near future.

NOT EVALUATED	DATA DEFICIENT	LEAST CONCERN	NEAR THREATENED	‹ VULNERABLE ›	ENDANGERED	CRITICALLY ENDANGERED	EXTINCT IN THE WILD	EXTINCT
NE	DD	LC	NT	VU	EN	CR	EW	EX

© José Vicente Rueda / CI Colombia

Hyalinobatrachium ibama, a species of Glass Frog, is listed as 'Vulnerable' on the IUCN Red List of Threatened Species™. It is found in Colombia in old-growth forests, in humid montane forests and near mountain streams. The distribution of this species is now severely fragmented.

The main threat facing Hyalinobatrachium ibama is considered to be habitat loss. The quality and extent of its forest habitat on the Eastern Cordillera of the Colombian Andes are currently declining due to cattle ranching and the illegal planting of crops. It is also considered to be threatened by the pollution associated with these agricultural activities, such as fumigation and fertilizer runoff.

Hyalinobatrachium ibama can be found in Santuario de Fauna y Flora Guanentá Alto Río Fonce Wildlife Sanctuary and in the protected Área Natural Única Los Estoraques, but more attention is needed to ensure that a greater proportion of its habitat is conserved.

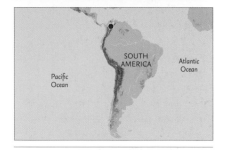

A species is Vulnerable when it faces a high risk of extinction in the wild, based on measurements of population size and/or geographic range and their trends in the past, present and/or future.

NOT EVALUATED	DATA DEFICIENT	LEAST CONCERN	NEAR THREATENED	⟨ VULNERABLE ⟩	ENDANGERED	CRITICALLY ENDANGERED	EXTINCT IN THE WILD	EXTINCT
NE	DD	LC	NT	VU	EN	CR	EW	EX

The **Sail-fin Lizard**, *Hydrosaurus pustulatus*, is listed as 'Vulnerable' on the IUCN Red List of Threatened Species™. This semi-aquatic reptile is found in the Philippines, and is as at home in water as it is in trees. These large creatures can exceed a metre in length, although much of this is tail.

The chief threat facing the Sail-fin Lizard is the destruction and degradation of its habitat. Its woodland habitat is being cleared for logging and agriculture, and its aquatic environment is often affected by water pollution and increased sedimentation due to erosion into waterways.

This erosion is caused by changing land use patterns, including increased logging. Hatchlings are particularly at risk from collection for the international and local pet trades, whilst the adults are hunted for consumption.

Sail-fin Lizards are fortunately found in many protected areas, but there is a need to better regulate the collection of this species from the wild, as populations are generally considered to be susceptible to overharvesting. There is also a need for improved regulations to prevent contamination from agrochemicals in waterways used by this species.

A species is **Vulnerable** when it faces a high risk of extinction in the wild, based on measurements of population size and/or geographic range and their trends in the past, present and/or future.

NOT EVALUATED	DATA DEFICIENT	LEAST CONCERN	NEAR THREATENED	VULNERABLE	⟨ ENDANGERED ⟩	CRITICALLY ENDANGERED	EXTINCT IN THE WILD	EXTINCT
NE	DD	LC	NT	VU	EN	CR	EW	EX

© D. Lin, California Academy of Sciences

The **São Tomé Giant Treefrog**, Hyperolius thomensis, is listed as 'Endangered' on the IUCN Red List of Threatened Species™. The largest member of its genus, this remarkable species is endemic to primary rainforest on the island of São Tomé in the Gulf of Guinea.

A notoriously difficult species to find, very little is known about the threats to the São Tomé Giant Treefrog. However, as a consequence of historically high levels of deforestation to make way for coffee and cocoa plantations, the species is now likely to be restricted to forest

fragments at altitudes above 800 metres. Urban and agricultural encroachment, livestock overgrazing and firewood collection are all thought to be continuing to threaten this species' habitat.

Although there are no known specific conservation measures in place for the São Tomé Giant Treefrog, much of its range is encompassed by the Obo National Park, offering this elusive species some sanctuary. However, with so little known about its ecology, population status or threats, further studies are urgently required.

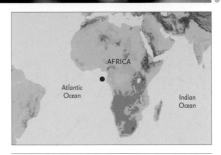

A species is Endangered when it faces a very high risk of extinction in the wild, based on measurements of population size and/or geographic range and their trends in the past, present and/or future.

Madagascar Giant Jumping Rat

Hypogeomys antimena

RED LIST

NOT EVALUATED	DATA DEFICIENT	LEAST CONCERN	NEAR THREATENED	VULNERABLE	‹ENDANGERED›	CRITICALLY ENDANGERED	EXTINCT IN THE WILD	EXTINCT
NE	DD	LC	NT	VU	EN	CR	EW	EX

The **Madagascar Giant Jumping Rat**, *Hypogeomys antimena*, is classified as 'Endangered' on the IUCN Red List of Threatened Species™. As their name suggests, they are adapted for jumping and can jump up almost a metre into the air as an evasive tactic to avoid predators. Through widespread loss of its dry forest habitat, the Giant Jumping Rat has declined dramatically and is now only found in a 300 km² patch of forest on the west coast of the island.

The main current threat to this species is further loss and degradation of its habitat through logging, clearance for agriculture and subsistence use of forest products. It is also thought to be susceptible to predation by dogs which are used by local hunters in the forest to catch other animals.

The Menabe Antimena protected area, encompassing the entire current range of the Giant Jumping Rat, was recently designated by the Government of Madagascar, providing hope that its remaining habitat can be protected. However, effective forest management and population monitoring are still needed to ensure the species is safeguarded.

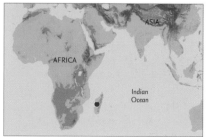

A species is **Endangered** when it faces a very high risk of extinction in the wild, based on measurements of population size and/or geographic range and their trends in the past, present and/or future.

Lesser Antillean Iguana

Iguana delicatissima

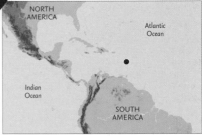

A species is **Endangered** when it faces a very high risk of extinction in the wild, based on measurements of population size and/or geographic range and their trends in the past, present and/or future.

© CHARLES KNAPP

The **Lesser Antillean Iguana**, *Iguana delicatissima*, is listed as 'Endangered' on the IUCN Red List of Threatened Species™. Once common throughout most of the northern Lesser Antilles of the Caribbean, this impressive lizard has been extirpated from several islands and is declining on most others.

Habitat clearance for agriculture was the main historical cause of the decline of this species. Now that tourism has taken over as the region's chief industry, coastal development has led to further habitat loss. Hunting of the Lesser Antillean Iguana was prevalent in the past and, although now illegal, is still common in some areas. Other significant threats include predation by feral mammals such as mongooses, cats and dogs, and hybridization with Common Iguanas (*Iguana iguana*).

The Lesser Antillean Iguana is legally protected throughout its range, but law enforcement is limited. It occurs in several nationally protected areas, but insufficient mitigation of existing threats lessens the conservation impact of these areas. Research on the species' population biology and ecology is ongoing, and captive breeding efforts are underway at the Jersey Wildlife Preservation Trust.

NOT EVALUATED	DATA DEFICIENT	LEAST CONCERN	NEAR THREATENED	⟨ VULNERABLE ⟩	ENDANGERED	CRITICALLY ENDANGERED	EXTINCT IN THE WILD	EXTINCT
NE	DD	LC	NT	VU	EN	CR	EW	EX

The **Usambara Drumming Grasshopper**, Ixalidium transiens, has not yet been officially evaluated for inclusion on the IUCN Red List of Threatened Species™, however it has a provisional assessment of 'Vulnerable'. It is endemic to the East Usambara Mountains in Tanzania and is a nocturnal species that dwells in the litter layer of evergreen submontane forests. It drums on the substrate with its hind legs to attract females.

The Usambara Drumming Grasshopper is a representative of a species-rich fauna found in the East Usambara Mountains. This mountain area is part of one of the global biodiversity hotspots, the Eastern Arc forests, which stretch from southeast Kenya to southwest Tanzania. The area, which is threatened by deforestation and habitat fragmentation, is known for its high endemism with 53% of the grasshopper species in this mountain block being endemic.

Due to the exceptionally high biodiversity found in the East Usambara Mountains, a lot of conservation activities have been implemented. Since 1993, the Tanzania Forest Conservation Group has been promoting forest conservation and attempting to establish wildlife corridors in order to reconnect the forest fragments.

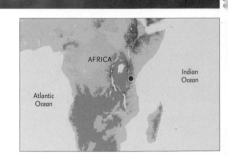

A species is Vulnerable when it faces a high risk of extinction in the wild, based on measurements of population size and/or geographic range and their trends in the past, present and/or future.

Guanaco

Lama guanicoe

RED LIST

NOT EVALUATED	DATA DEFICIENT	LEAST CONCERN	NEAR THREATENED	VULNERABLE	ENDANGERED	CRITICALLY ENDANGERED	EXTINCT IN THE WILD	EXTINCT
NE	DD	LC	NT	VU	EN	CR	EW	EX

© JAVIER ETCHEVERRY

The **Guanaco**, *Lama guanicoe*, is listed as 'Least Concern' on the IUCN Red List of Threatened Species™. Being the wild relative of the domestic Llama, this elegant animal has a widespread but rather fragmented distribution across much of South America.

Although still relatively widespread and abundant, the Guanaco now occupies only around 40 percent of its original range, and many populations have become quite isolated. The main threats to the species include overhunting and poaching, habitat degradation (often due to overgrazing and drought), competition with domestic livestock, and the use of barbed wire fences.

The Guanaco is considered dependent on conservation measures for its long-term survival. Guanacos are legally protected in many areas, and international trade in the species is regulated by CITES, but illegal hunting still persists. In Argentina, home to 90% of the Guanaco population, a national management plan is in place, while sustainable use programmes, such as live-shearing initiatives as an alternative to killing the animals for their fibre, may provide incentives to protect wild Guanaco populations.

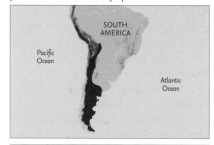

A species is **Least Concern** when it does not qualify for Critically Endangered, Endangered, Vulnerable or Near Threatened. Widespread and abundant species are included in this category.

Porbeagle

Lamna nasus

The **Porbeagle**, *Lamna nasus*, is listed as 'Vulnerable' on the IUCN Red List of Threatened Species™. It is found worldwide in temperate and cold-temperate waters. Subpopulations in the Northeast Atlantic and Mediterranean are classified as 'Critically Endangered', and that of the Northwest Atlantic as 'Endangered'.

The greatest threat to this shark is unsustainable commercial fishing. A low reproductive rate and high commercial value, both in target and incidental fisheries, make the Porbeagle highly vulnerable to overexploitation, and populations in the North Atlantic have been seriously depleted, while those in the Mediterranean have virtually disappeared. Little information is available from the southern oceans, where its population status is unknown.

Catches of Porbeagle are regulated in the EU and New Zealand, and the species is included in fishery management plans in Canada and the USA. However, the international trade in Porbeagle meat that drives many fisheries is currently unregulated, and a proposal to list the species on Appendix II of CITES, which makes international trade without a permit illegal, was rejected earlier this year.

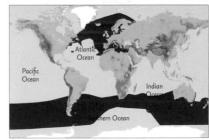

A species is **Vulnerable** when it faces a high risk of extinction in the wild, based on measurements of population size and/or geographic range and their trends in the past, present and/or future.

São Tomé Fiscal

Lanius newtoni

NOT EVALUATED	DATA DEFICIENT	LEAST CONCERN	NEAR THREATENED	VULNERABLE	ENDANGERED	CRITICALLY ENDANGERED	EXTINCT IN THE WILD	EXTINCT
NE	DD	LC	NT	VU	EN	CR	EW	EX

© Martim Melo

The **São Tomé Fiscal**, *Lanius newtoni*, is listed as 'Critically Endangered' on the IUCN Red List of Threatened Species™. Endemic to the island of São Tomé in Africa's Gulf of Guinea, this long-tailed shrike was not observed in the wild between 1928 and 1990, raising concerns that it might have gone extinct.

Historically, large areas of São Tomé's forests were cleared for cocoa and coffee plantations. This habitat loss continues in places today, with clearance for small farms and road-building. The São Tomé Fiscal is now restricted to small areas of remaining primary forest, and its population is believed to be tiny. Introduced predators such as rats, stoats and monkeys may also be a threat.

São Tomé's remaining primary forest is protected in Obo National Park, although poor law enforcement and a lack of information on the São Tomé Fiscal make the benefits to this species unclear. Further research into the species will be vital, while its listing as a protected species, together with the legal protection and active management of its habitat, will also be important in its conservation.

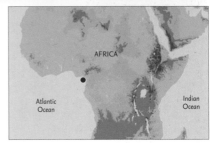

A species is *Critically Endangered* when it faces an extremely high risk of extinction in the wild, based on measurements of population size and/or geographic range and their trends in the past, present and/or future.

Laotian Rock Rat

Laonastes aenigmamus

NOT EVALUATED	DATA DEFICIENT	LEAST CONCERN	NEAR THREATENED	VULNERABLE	‹ ENDANGERED ›	CRITICALLY ENDANGERED	EXTINCT IN THE WILD	EXTINCT
NE	DD	LC	NT	VU	EN	CR	EW	EX

The **Laotian Rock Rat**, *Laonastes aenigmamus*, is listed as 'Endangered' on the IUCN Red List of Threatened Species™. This recently described species is the sole surviving member of the Diatomyidae family, an ancient group of rodents previously considered to have been extinct for some 11 million years.

While the status of the Laotian Rock Rat is not entirely clear, it is thought to be threatened by hunting and habitat loss. Hunting of the Laotian Rock Rat is prevalent throughout its range, and as the species was identified by specimens from a bushmeat market, there is concern that this hunting is unsustainable. The species' karst habitat is relatively safe from large-scale destruction; however, more accessible areas are vulnerable to degradation from logging and firewood collection.

The Laotian Rock Rat is afforded some sanctuary in the Khammouan Limestone National Biodiversity Conservation Area. However, with very little known about this enigmatic species, there is a need for further surveys to evaluate its species population status and threats.

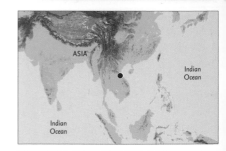

A species is *Endangered* when it faces a very high risk of extinction in the wild, based on measurements of population size and/or geographic range and their trends in the past, present and/or future.

Northern Hairy-nosed Wombat

RED LIST

Lasiorhinus krefftii

NOT EVALUATED	DATA DEFICIENT	LEAST CONCERN	NEAR THREATENED	VULNERABLE	ENDANGERED	CRITICALLY ENDANGERED	EXTINCT IN THE WILD	EXTINCT
NE	DD	LC	NT	VU	EN	CR	EW	EX

The **Northern Hairy-nosed Wombat**, *Lasiorhinus krefftii*, is listed as 'Critically Endangered' on the IUCN Red List of Threatened Species™. One of the world's rarest mammals, it is now only found at a single location in Queensland, Australia, and the total population numbers only around 138 individuals (2007 estimate).

This species may have already been uncommon before European settlers arrived, but due to a combination of drought and competition with grazing livestock, the population decline accelerated. Its small colony is further threatened by unpredictable environmental effects, loss of genetic diversity and exotic buffel grass taking over its natural habitat.

The entire range of the Northern Hairy-nosed Wombat is encompassed by the Epping Forest National Park. A recovery plan has been implemented and, as a result, a 20 km long dingo and cattle exclusion fence has been built, its habitat is managed, and kangaroo numbers are monitored. A second colony in the wild has recently been established, with the transfer of a small number of wombats to a reserve in south-central Queensland, with promising results so far. There are also plans to develop a captive breeding programme.

© Darren Jew, Department of Environment and Resource Management

A species is Critically Endangered when it faces an extremely high risk of extinction in the wild, based on measurements of population size and/or geographic range and their trends in the past, present and/or future.

Sulawesi Coelacanth

Latimeria menadoensis

NOT EVALUATED	DATA DEFICIENT	LEAST CONCERN	NEAR THREATENED	⟨ VULNERABLE ⟩	ENDANGERED	CRITICALLY ENDANGERED	EXTINCT IN THE WILD	EXTINCT
NE	DD	LC	NT	VU	EN	CR	EW	EX

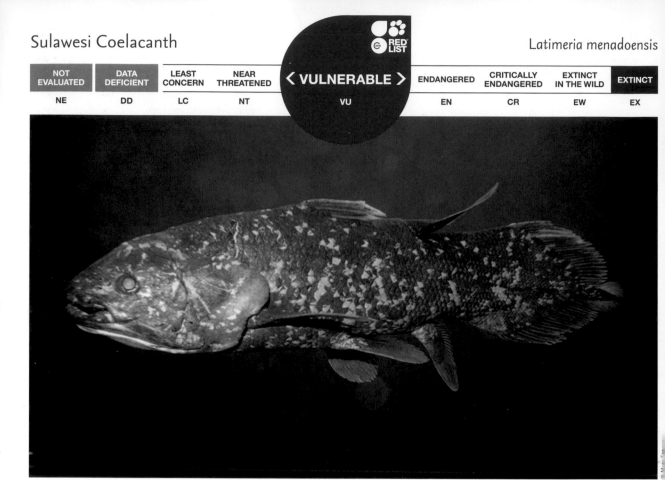

The **Sulawesi Coelacanth**, *Latimeria menadoensis*, is classified as 'Vulnerable' on the IUCN Red List of Threatened Species™. Coelacanths were thought to be extinct; however, in 1997 a discovery of *Latimeria menadoensis* was made off the northern Sulawesi coast, Indonesia.

The Sulawesi Coelacanth can be found at depths of about 150–200m on rocky slopes or in caves. Not much is known about them and few observations have been made in the wild. The major threats to the species include by-catch in deep-water shark nets, and hook-and-line fishing for deep-water groupers. This species has also recently been targeted for live capture and exhibition in large aquaria, although no individuals have been successfully kept alive.

Current conservation efforts to preserve these 'living fossils' are becoming more stringent. The Sulawesi Coelacanth is currently protected under Indonesian fishing regulations with the recent ban on shark nets in the Bunakem National Park, as well as international trade protection under CITES. Continued research will help to provide more information on this species to establish future protection measures.

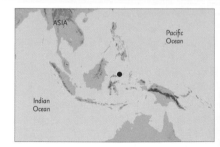

A species is Vulnerable when it faces a high risk of extinction in the wild, based on measurements of population size and/or geographic range and their trends in the past, present and/or future.

NOT EVALUATED	DATA DEFICIENT	LEAST CONCERN	NEAR THREATENED	⟨ VULNERABLE ⟩	ENDANGERED	CRITICALLY ENDANGERED	EXTINCT IN THE WILD	EXTINCT
NE	DD	LC	NT	VU	EN	CR	EW	EX

© Nik Cole - Durrell Wildlife Conservation Trust

The **Round Island Skink**, *Leiolopisma telfairii*, is classified as 'Vulnerable' on the IUCN Red List of Threatened Species™. It is the largest surviving lizard of a diverse and unique community of reptiles that once dominated the Indian Ocean island of Mauritius. Following the arrival of people on Mauritius in the 16th century, the skink became restricted to the remote northern island, Round Island. At only 2.15 km², this island currently supports an estimated 26,000 individuals.

The introduction of mammalian predators, such as rats, which decimate island reptile populations, is the main threat to the survival of this species. Round Island is one of only a few islands in the Indian Ocean that has never been invaded by rats. For this reason, the Round Island Skink still survives.

Since the removal of introduced rabbits and goats from Round Island, and aided by extensive habitat restoration, the skink population continues to grow. The eradication of introduced mammals from other islands around Mauritius has recently permitted conservationists to reintroduce the skink back to islands within its former range.

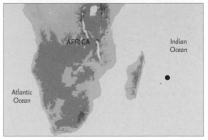

A species is **Vulnerable** when it faces a high risk of extinction in the wild, based on measurements of population size and/or geographic range and their trends in the past, present and/or future.

Pampas Cat

Leopardus colocolo

NOT EVALUATED	DATA DEFICIENT	LEAST CONCERN	NEAR THREATENED	VULNERABLE	ENDANGERED	CRITICALLY ENDANGERED	EXTINCT IN THE WILD	EXTINCT
NE	DD	LC	NT	VU	EN	CR	EW	EX

The **Pampas Cat**, Leopardus colocolo, is listed as 'Near Threatened' on the IUCN Red List of Threatened Species™. It occurs in central Brazil, and from the Peruvian Andes to southwestern South America, where, as the name suggests, it is typically associated with pampas grasslands, although it can be found in a very broad range of habitats.

Until international trade ceased in 1987, the Pampas Cat was hunted extensively for pelts. Today, habitat destruction is likely to be the main threat, with much of the pampas grasslands having been converted to agriculture or heavily grazed.

The Pampas Cat is also killed in retaliation for taking poultry, and for traditional cultural purposes in some areas.

Hunting of the Pampas Cat is prohibited in a number of countries, although the species currently receives no legal protection in Ecuador. The Pampas Cat occurs within a number of protected areas in Argentina, and a few protected areas within the Brazilian savannas (Cerrado). Further research into its ecology, taxonomy and distribution has been recommended to help in its conservation.

A species is **Near Threatened** when it does not qualify for Critically Endangered, Endangered or Vulnerable now, but is close to qualifying for or is likely to qualify for a threatened category in the near future.

Little Mountain Palm

Lepidorrhachis mooreana

NOT EVALUATED	DATA DEFICIENT	LEAST CONCERN	NEAR THREATENED	VULNERABLE	⟨ ENDANGERED ⟩	CRITICALLY ENDANGERED	EXTINCT IN THE WILD	EXTINCT
NE	DD	LC	NT	VU	EN	CR	EW	EX

The **Little Mountain Palm**, Lepidorrhachis mooreana, is not currently listed on the IUCN Red List of Threatened Species™, however a preliminary evaluation has classified it as 'Endangered'. A relatively short, solitary palm, it is restricted to the summit of Mt. Gower on Lord Howe Island, in the southwest Pacific.

This endemic palm occurs in dwarf mossy forest at elevations above 750 metres, in an area totalling less than 0.5 km². It is under threat from introduced rats which eat the fruits, preventing regeneration, as well as from invasive weeds and pathogens, and damage to the fragile soil and vegetation by human visitors. Climate change may also affect cloud cover, with potentially catastrophic effects on the forest.

The entire distribution of this species falls within a reserve. A number of conservation efforts are underway on Lord Howe Island, including controls on tourist access, a comprehensive weed strategy, and research into the Little Mountain Palm. This species would also benefit greatly from a complete rat eradication programme.

© William J. Baker / RBG Kew

A species is Endangered when it faces a very high risk of extinction in the wild, based on measurements of population size and/or geographic range and their trends in the past, present and/or future.

Sarawak Land Crab

Lepidothelphusa cognetti

NOT EVALUATED	DATA DEFICIENT	LEAST CONCERN	NEAR THREATENED	VULNERABLE	⟨ENDANGERED⟩	CRITICALLY ENDANGERED	EXTINCT IN THE WILD	EXTINCT
NE	DD	LC	NT	VU	EN	CR	EW	EX

A species is **Endangered** when it faces a *very high risk of extinction in the wild*, based on measurements of population size and/or geographic range and their trends in the past, present and/or future.

The **Sarawak Land Crab**, *Lepidothelphusa cognetti*, is listed as 'Endangered' on the IUCN Red List of Threatened Species™. These long-legged, black and white, semi-terrestrial crabs are endemic to Sarawak, East Malaysia, on the island of Borneo. So far, this species has only been found only in two locations, both associated with sandstone rocks, in western Sarawak on Mount Penrissen, and Bau, a gold-mining town.

The main threats to the Sarawak Land Crab are habitat loss and degradation, primarily as a result of logging. For example, much of the suitable habitat within its range around Mount Penrissen has recently been converted into a golf course and to cultivated land. An additional threat from deforestation is the resultant sedimentation of the streams in which it lives.

The Sarawak Land Crab is not found in a protected area, and this raises questions about the long-term survival of this threatened and unique species.

Northern Sportive Lemur

NOT EVALUATED	DATA DEFICIENT	LEAST CONCERN	NEAR THREATENED	VULNERABLE	ENDANGERED	CRITICALLY ENDANGERED	EXTINCT IN THE WILD	EXTINCT
NE	DD	LC	NT	VU	EN	CR	EW	EX

The **Northern Sportive Lemur,** Lepilemur septentrionalis, is listed as 'Critically Endangered' on the IUCN Red List of Threatened Species™. This small, arboreal primate is restricted to the very northernmost parts of Madagascar, and has a fragmented population of probably no more than 100 individuals.

Like much of the native fauna of Madagascar, habitat loss presents the greatest threat to the Northern Sportive Lemur, with the felling of trees for charcoal production greatly reducing the area of forest available to this species. In spite of being officially protected, this lemur is also hunted for food.

With the entire population occurring on unprotected land, this species has an extremely uncertain future. There is some hope that further survey work will reveal the presence of previously unrecorded subpopulations in other forest patches.

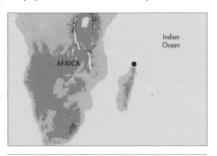

A species is Critically Endangered when it faces an extremely high risk of extinction in the wild, based on measurements of population size and/or geographic range and their trends in the past, present and/or future.

© R.A. MITTERMEIER

NOT EVALUATED	DATA DEFICIENT	LEAST CONCERN	NEAR THREATENED	VULNERABLE	ENDANGERED	CRITICALLY ENDANGERED	EXTINCT IN THE WILD	EXTINCT
NE	DD	LC	NT	VU	EN	CR	EW	EX

The **Mountain Chicken**, Leptodactylus fallax, is classified as 'Critically Endangered' on the IUCN Red List of Threatened Species™. One of the world's most threatened frogs, the oddly named Mountain Chicken is so called because its meat is said to taste like chicken. It is also one of the largest frogs in the world, with adult females growing up to remarkable lengths of 21 centimetres. It currently occurs on Dominica and Montserrat in the Eastern Caribbean.

The two remaining populations of the Mountain Chicken declined as a result of hunting, volcanic eruptions, introduced predators and habitat loss. More recently, the arrival of chytrid fungus on both islands has led to catastrophic declines and near extinction of the species.

Intensive conservation of this species began in 1999 when the first Montserratian frogs were taken into captivity at Durrell Wildlife Conservation Trust for breeding. The Zoological Society of London became involved with the species in Dominica in 2004. Local wildlife agencies and other European zoos are now involved in both in situ and ex situ conservation efforts, and the first releases of captive-bred offspring are planned for the near future.

A species is **Critically Endangered** when it faces an extremely high risk of extinction in the wild, based on measurements of population size and/or geographic range and their trends in the past, present and/or future.

Tehuantepec Jackrabbit

Lepus flavigularis

NOT EVALUATED	DATA DEFICIENT	LEAST CONCERN	NEAR THREATENED	VULNERABLE	⟨ ENDANGERED ⟩	CRITICALLY ENDANGERED	EXTINCT IN THE WILD	EXTINCT
NE	DD	LC	NT	VU	EN	CR	EW	EX

The **Tehuantepec Jackrabbit**, *Lepus flavigularis*, is listed as 'Endangered' on the IUCN Red List of Threatened Species™. Found only in the state of Oaxaca in southern Mexico, where it has been reduced to just four small, isolated populations, the Tehuantepec Jackrabbit is considered the most endangered hare species in the world.

The main threats to this species are hunting for food and sport, ongoing habitat loss, conversion of grasslands to agriculture, pasture or settlement, habitat degradation by human-induced fires, and the spread of exotic grasses. The small remaining population, estimated at fewer than 1,000 individuals, also has low genetic variation, which may potentially lead to problems associated with inbreeding.

Although listed as 'Critically Endangered' within Mexico, conservation laws are not properly enforced, and the areas the species inhabits receive no formal protection. Further research is needed on the Tehuantepec Jackrabbit, and captive breeding has also been suggested. However, without better control of hunting and urgent action to protect remaining habitat, this highly threatened mammal faces an uncertain future.

© ARTURO CARRILLO REYES

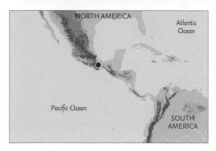

A species is **Endangered** when it faces a very high risk of extinction in the wild, based on measurements of population size and/or geographic range and their trends in the past, present and/or future.

Patagonian Opossum

Lestodelphys halli

The **Patagonian Opossum**, Lestodelphys halli, is listed as 'Least Concern' on the IUCN Red List of Threatened Species™. The sole member of the genus Lestodelphys, this opossum occurs farther south in the region than any other living marsupial, and is one of the least-known representatives of this group. It predominantly inhabits the dry and cold environments of Patagonia and the lower Monte Desert, Argentina, where it feeds mostly on mice and arthropods.

Until recently, the species was known from fewer than 10 specimens distributed from approximately 33° to 48° S, but remains recovered from owl pellets have increased the number of localities to more than 90. The main threat to the Patagonian Opossum is habitat loss due to increased desertification of Patagonia.

The Patagonian Opossum is found in a number of protected areas across its range. With so little known about its exact geographic distribution, ecology and natural history, further research is required to determine the exact status of this species.

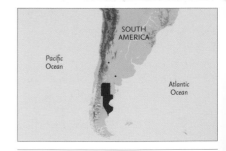

A species is **Least Concern** when it does not qualify for Critically Endangered, Endangered, Vulnerable or Near Threatened. Widespread and abundant species are included in this category.

Purple Skimmer

Libellula jesseana

© Dennis Paulson

The **Purple Skimmer**, *Libellula jesseana*, is classified as 'Vulnerable' on the IUCN Red List of Threatened Species™. This species has a highly restricted range and occurs only in parts of Florida where there are clean, sandy lakes with very little vegetation. The males perch in tall grasses and fly across the lake when fiercely competing for territories. Females forage in clearings in the nearby woods and larvae live in silty sediment on the lake bottom.

Research has found that this species is more common and widespread than originally thought.

However, the Purple Skimmer remains threatened by the rapid development that has occurred around Florida's beautiful lakes. This development has degraded this dragonfly's habitat through water pollution and ground water depletion.

One population is protected within a state park and another within a national forest. The best-known population of the Purple Skimmer is protected in Gold Head Branch State Park, but most other lakes of historical occurrence are not protected. Further monitoring is essential for the future wellbeing of this species of dragonfly.

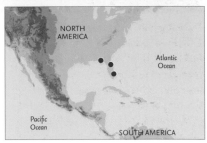

A species is Vulnerable when it faces a high risk of extinction in the wild, based on measurements of population size and/or geographic range and their trends in the past, present and/or future.

NOT EVALUATED	DATA DEFICIENT	LEAST CONCERN	NEAR THREATENED	‹ VULNERABLE ›	ENDANGERED	CRITICALLY ENDANGERED	EXTINCT IN THE WILD	EXTINCT
NE	DD	LC	NT	VU	EN	CR	EW	EX

© DI STEPHENSON

The **Aquatic Tenrec**, *Limnogale mergulus*, is listed as 'Vulnerable' on the IUCN Red List of Threatened Species™. This nocturnal, semiaquatic mammal, also known as the Web-footed Tenrec, is endemic to the eastern humid forests and central highlands of Madagascar. It feeds on aquatic invertebrates and may be dependent on the availability of permanent, clean and fast-flowing water, where its prey thrives.

Owing to its specific habitat requirements, the Aquatic Tenrec is threatened by soil erosion and siltation caused by deforestation. Agricultural expansion is fragmenting the upland forests, thereby isolating fast-flowing riverine habitat. Individuals are also sometimes killed in fish traps.

This species is only known from two protected areas: Ranomafana National Park and the Andringitra National Park. The prevention of erosion and sedimentation is of paramount importance for its conservation. Therefore, Aquatic Tenrec habitat needs to be protected from sedimentation wherever possible, for instance by inclusion of forested catchments in the protected areas network, by effective terracing of agricultural fields, and by maintenance of vegetated riparian zones.

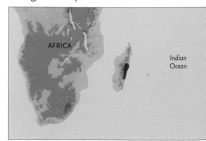

AFRICA

Indian Ocean

Pigeon Beak

Lotus maculatus

RED LIST

NOT EVALUATED	DATA DEFICIENT	LEAST CONCERN	NEAR THREATENED	VULNERABLE	ENDANGERED	CRITICALLY ENDANGERED	EXTINCT IN THE WILD	EXTINCT
NE	DD	LC	NT	VU	EN	CR	EW	EX

© Klaus Nickel

Pigeon Beak, *Lotus maculatus*, is classified as 'Critically Endangered' on the Red List of Spanish Vascular Flora. It is exclusively known from one site on the island of Tenerife (Canary Islands, Spain). It is a creeping plant with beautiful and abundant yellow-red flowers, which has been used as a carpet plant for gardens. Recent field studies have confirmed the extinction of the species in a previously known area, so now only 10 individuals remain on the northern coast of Tenerife.

The main threats to this species are related to human activities such as development, tourism, the collection of wild plants and the presence of herbivores. A lack of pollinators may also pose a potential threat to the Pigeon Beak.

Fortunately, the species is covered by legislation in Spain at both the regional and national level, as well as by the Bern Convention. It is also included in a recovery plan; some reintroductions have been made in several locations on the island, and seeds have been stored in gene banks.

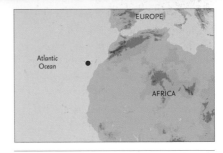

A species is Critically Endangered when it faces an extremely high risk of extinction in the wild, based on measurements of population size and/or geographic range and their trends in the past, present and/or future.

African Elephant

Loxodonta africana

The **African Elephant**, Loxodonta africana, is listed as 'Vulnerable' on the IUCN Red List of Threatened Species™. It is the largest living terrestrial animal and is found across much of sub-Saharan Africa, but its populations are becoming increasingly fragmented.

Historically, poaching for ivory and meat has been the main cause of the African Elephant's decline. However, while illegal hunting remains a significant threat in some areas, habitat loss and fragmentation due to human population expansion and rapid land conversion pose major challenges over much of the range, and increase the incidence of human-elephant conflict.

The African Elephant receives various degrees of legal protection throughout its range, and international trade in elephant ivory is controlled under CITES. Sport hunting is permitted in some countries, and several countries have CITES export quotas for elephant trophies. Effective management and conservation has been successful in increasing elephant numbers in southern and eastern Africa, but varying approaches are needed for the specific problems facing this charismatic species in different countries and regions across its range.

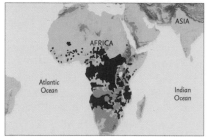

A species is Vulnerable when it faces a high risk of extinction in the wild, based on measurements of population size and/or geographic range and their trends in the past, present and/or future.

African Wild Dog

Lycaon pictus

RED LIST

NOT EVALUATED	DATA DEFICIENT	LEAST CONCERN	NEAR THREATENED	VULNERABLE	‹ ENDANGERED ›	CRITICALLY ENDANGERED	EXTINCT IN THE WILD	EXTINCT
NE	DD	LC	NT	VU	EN	CR	EW	EX

© Patricia Medici

The **African Wild Dog**, Lycaon pictus, is listed as 'Endangered' on the IUCN Red List of Threatened Species™. This species is limited to only a portion of its historical distribution, with populations eradicated from West Africa and greatly reduced in central and northeast Africa. They are rarely seen due to their low population densities.

Wild Dogs experience high mortality in comparison with other large carnivore species. While competition from other predators, primarily lions, is the principal cause of natural mortality in adults, more than half of all deaths are due to human activity. Wild Dogs ranging outside of protected areas encounter high-speed vehicles, guns, snares and poisons, as well as domestic dogs, which represent reservoirs of potentially lethal diseases (rabies and canine distemper).

The establishment of protected areas, as well as conservancies on private and communal land, has decreased contact between Wild Dogs, people and domestic dogs. There are also efforts to work with local people to reduce deliberate killing of the dogs in and around these protected areas. Establishing effective techniques for protecting small populations from serious infections is also a priority.

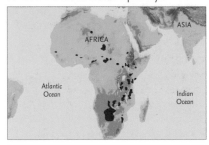

A species is Endangered when it faces a very high risk of extinction in the wild, based on measurements of population size and/or geographic range and their trends in the past, present and/or future.

Madagascan Land Crab

Madagapotamon humberti

The **Madagascan Land Crab**, *Madagapotamon humberti*, is listed as 'Vulnerable' on the IUCN Red List of Threatened Species™. This long-legged, brightly coloured, air-breathing land crab is restricted to limestone formations, and requires a specialized habitat of water-filled crevices in limestone.

The Ankarana Massif in northern Madagascar is an undeveloped region of the island with vast limestone formations that include caves, underground rivers, and forested valleys. The Madagascan Land Crab spends its time within deep fissures in the limestone during the dry season, and emerges to feed in the nearby forests when the rainy season begins. The specialized habitat of this species limits its distribution to this part of the island, and it is endemic to Madagascar.

Despite the fact that several specimens have been collected recently from different localities, the long-term threats to the survival of this species remain, and include habitat disturbance and pollution. In addition, only part of its range lies within a protected area, so the long-term survival of this vulnerable and ecologically unusual species could be at risk.

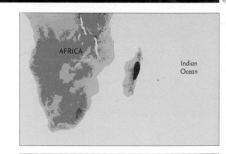

A species is Vulnerable when it faces a high risk of extinction in the wild, based on measurements of population size and/or geographic range and their trends in the past, present and/or future.

Biznaguita

NOT EVALUATED	DATA DEFICIENT	LEAST CONCERN	NEAR THREATENED	VULNERABLE	ENDANGERED	CRITICALLY ENDANGERED	EXTINCT IN THE WILD	EXTINCT
NE	DD	LC	NT	VU	EN	CR	EW	EX

Biznaguita, *Mammillaria herrerae*, is listed as 'Critically Endangered' on the IUCN Red List of Threatened Species™. Endemic to Querétaro, Mexico, this unusual cactus has the appearance of a small golf ball, its white spines forming an almost lacy-looking, spherical surface.

Biznaguita is under serious threat from illegal collection and, as a result, has declined by an estimated 95 percent over the past 20 years. The wild population is now reduced to approximtely 50 individuals, growing in an area of just over one square kilometre. A nearby commercial cactus nursery has almost totally stripped the site, whilst local children often collect the plants and offer them to visitors. There is also a small amount of residential development in the area, introducing a further potential threat.

Although listed on Appendix II of CITES, meaning international trade in this cactus should be carefully controlled, greater enforcement of these regulations is needed. Biznaguita has long been propagated around the world, which may provide hope for the species' survival should the wild population be lost.

© Rolando Bárcenas

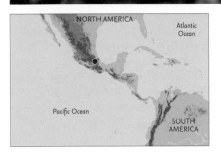

A species is **Critically Endangered** when it faces an extremely high risk of extinction in the wild, based on measurements of population size and/or geographic range and their trends in the past, present and/or future.

NOT EVALUATED	DATA DEFICIENT	LEAST CONCERN	NEAR THREATENED	VULNERABLE	ENDANGERED	CRITICALLY ENDANGERED	EXTINCT IN THE WILD	EXTINCT
NE	DD	LC	NT	VU	EN	CR	EW	EX

Manglietia sinica, also known as Magnolia sinica, is listed as 'Critically Endangered' on the IUCN Red List of Threatened Species™. This magnolia species occurs on forested slopes in southeast Yunnan, China, where it is known from just a single, tiny population.

Manglietia sinica is considered to be one of the most threatened magnolias in the world. Surveys in 2005–2006 found just 10 mature wild individuals, although subsequent work found an additional 40 trees, bringing the total wild population to 50 individuals. The species remains under threat from cutting, habitat destruction and poor natural regeneration.

A number of projects are now underway to restore and protect this rare plant. Several thousand saplings are currently held in nurseries, and individuals are gradually being replanted in the wild. The Global Trees Campaign has been working in partnership with various organisations to conserve this species. These efforts have helped to develop support for Wenshan National Nature Reserve, where the saplings are being planted, and enhanced the reserve's long-term future, giving added protection to this and other threatened species.

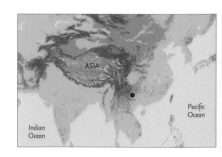

Cowan's Mantella

Mantella cowani

NOT EVALUATED	DATA DEFICIENT	LEAST CONCERN	NEAR THREATENED	VULNERABLE	ENDANGERED	CRITICALLY ENDANGERED	EXTINCT IN THE WILD	EXTINCT
NE	DD	LC	NT	VU	EN	CR	EW	EX

© Franco Andreone

Cowan's Mantella (or the Harlequin Mantella), *Mantella cowani*, is now listed as 'Critically Endangered' on the IUCN Red List of Threatened Species™. It is a small venomous frog found in high altitude evergreen forests that run along rivers and streams in eastern central Madagascar. It can also be found in high altitude grassland savannas during the rainy season.

This species has had to endure massive exploitation for the international pet trade as well as deforestation, which has left its habitat fragmented and, in some areas, destroyed. Its remaining forest habitat is under threat from increasing subsistence agriculture, timber extraction, charcoal manufacture, livestock grazing, fires and human settlements. As a result, Cowan's Mantella now only occupies a space of 10km², and its population is estimated to have decreased by more than 80% over the last 15 years.

Work continues on protecting Cowan's Mantella's remaining habitats and on implementing the export ban, at least until populations of this species can recover.

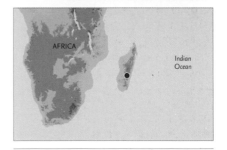

A species is **Critically Endangered** when it faces an extremely high risk of extinction in the wild, based on measurements of population size and/or geographic range and their trends in the past, present and/or future.

Freshwater Pearl Mussel

Margaritifera margaritifera

NOT EVALUATED	DATA DEFICIENT	LEAST CONCERN	NEAR THREATENED	VULNERABLE	⟨ENDANGERED⟩	CRITICALLY ENDANGERED	EXTINCT IN THE WILD	EXTINCT
NE	DD	LC	NT	VU	EN	CR	EW	EX

The **Freshwater Pearl Mussel**, *Margaritifera margaritifera*, is listed as 'Endangered' on the IUCN Red List of Threatened Species™. Its distribution stretches from northern Spain through France, the United Kingdom and central Europe, to Scandinavia and northern Russia. Individuals develop very slowly; in Sweden, recently discovered specimens are up to 280 years old.

The presence of reproducing populations of the Freshwater Pearl Mussel indicates a high quality of freshwater ecosystems. The main threats to this species are degradation of its habitat (particularly intensification of land use leading to sedimentation), river regulation and nutrient pollution.

The majority of European populations have mainly adult mussels remaining, with unsustainable levels of reproduction. This is due to the continued loss of juvenile mussels from a lack of oxygen reaching the riverbed gravels, where they are completely buried for their first five years.

Catchment management to restore riverbed habitat conditions is considered to be vital to the prevention of the extinction of this species. As an indicator species, it can be seen as a model for designing management plans for freshwater ecosystems.

A species is Endangered when it faces a very high risk of extinction in the wild, based on measurements of population size and/or geographic range and their trends in the past, present and/or future.

Vancouver Island Marmot

NOT EVALUATED	DATA DEFICIENT	LEAST CONCERN	NEAR THREATENED	VULNERABLE	ENDANGERED	CRITICALLY ENDANGERED	EXTINCT IN THE WILD	EXTINCT
NE	DD	LC	NT	VU	EN	CR	EW	EX

The **Vancouver Island Marmot**, *Marmota vancouverensis*, is listed as 'Critically Endangered' on the IUCN Red List of Threatened Species™. One of the rarest mammals in North America, this species is only found on Vancouver Island, Canada. Its total population was estimated at fewer than 130 individuals in 2004, most of which were in captivity.

Natural successional processes, such as the advancement of tree growth in alpine meadows, may be responsible for this marmot's original scarcity. However, the main causes of its recent decline may be habitat disruption through logging, and high predation rates associated with forestry practices and changes in the numbers of predators. With such a small population, any predation on the Vancouver Island Marmot may represent a significant threat.

The Vancouver Island Marmot is legally protected, and a recovery plan is in place for the species. A captive breeding programme is underway, and reintroductions have been planned. Detailed research is needed to better understand aspects of the marmot's dispersal, survivorship, requirements for hibernation sites, and its reproduction in natural and logged habitats.

© FLETCHER & BAYLIS

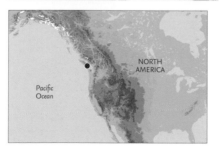

A species is **Critically Endangered** when it faces an extremely high risk of extinction in the wild, based on measurements of population size and/or geographic range and their trends in the past, present and/or future.

Waterclover

Marsilea azorica

NOT EVALUATED	DATA DEFICIENT	LEAST CONCERN	NEAR THREATENED	VULNERABLE	ENDANGERED	CRITICALLY ENDANGERED	EXTINCT IN THE WILD	EXTINCT
NE	DD	LC	NT	VU	EN	CR	EW	EX

Waterclover, *Marsilea azorica*, is listed as 'Critically Endangered' on the IUCN Red List of Threatened Species™. It is a small fern of 5–15 cm in height, with four beautiful leaflets. Waterclover only grows in a small seasonal puddle on the Terceira Island of the Azores Archipelago.

The main threats to the Waterclover are related to agricultural development, and the contamination and degradation of its habitat by the establishment of exotic species. The isolation of the population and its limited dispersal make this small fern highly vulnerable to genetic impoverishment.

Waterclover is included on the Bern Convention and on the EU Habitats Directive, the aim of which is to contribute towards ensuring the survival of biodiversity through the conservation of natural habitats and of wild fauna and flora within the European Union. However, urgent conservation actions are required to save this species by conserving its habitat, especially through banning the use of fertilizers and chemical products around the small pool where it lives, and fencing it off in order to prevent livestock access.

A species is **Critically Endangered** when it faces an extremely high risk of extinction in the wild, based on measurements of population size and/or geographic range and their trends in the past, present and/or future.

NOT EVALUATED	DATA DEFICIENT	LEAST CONCERN	NEAR THREATENED	VULNERABLE	ENDANGERED	CRITICALLY ENDANGERED	EXTINCT IN THE WILD	EXTINCT
NE	DD	LC	NT	VU	EN	CR	EW	EX

© WILLIAM J. BAKER / RBG, KEW

Medemia argun is listed as 'Critically Endangered' on the IUCN Red List of Threatened Species™. This dramatic, single-stemmed palm occurs as scattered individuals and populations in the Nubian Desert of southern Egypt and northern Sudan. Although first described from fruits collected from Egyptian tombs, it wasn't until 1845 that it was described from living material. For many years it was considered Extinct in the Wild, until its rediscovery in 1995.

The rarity of this palm in current times is paradoxical, given its frequent occurrence in ancient tombs far to the north. It seems likely that it once had a broader distribution than it has today. It is known that the Sahara has become more arid over this time frame, and therefore the range of *Medemia argun* may be shrinking as a result of long-term climate change.

A comprehensive census of unexplored potential sites is underway in Egypt. In addition, a small nursery has been established in Aswan as an ex situ resource, while seedlings have been distributed as part of a local awareness campaign.

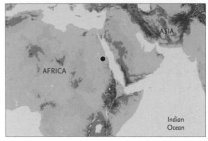

A species is Critically Endangered when it faces an extremely high risk of extinction in the wild, based on measurements of population size and/or geographic range and their trends in the past, present and/or future.

NOT EVALUATED	DATA DEFICIENT	LEAST CONCERN	NEAR THREATENED	VULNERABLE	ENDANGERED	CRITICALLY ENDANGERED	EXTINCT IN THE WILD	EXTINCT
NE	DD	LC	NT	VU	EN	CR	EW	EX

The Brazilian land snail, *Megalobulimus grandis*, is classified as 'Critically Endangered' on the IUCN Red List of Threatened Species™. It is one of the largest land snails found in southeastern Brazil, where it is endemic to the Atlantic Rainforest in the north coastal region of São Paulo State. Although little is known about the biology of this species, it is presumed to live on the forest floor amongst the leaf litter.

All the Atlantic Forest occurring in São Paulo State has been declared to be under special protection. However, the forest is now extremely fragmented, with only 7% of the original forest area remaining following centuries of deforestation, which poses a serious threat to this species.

The invertebrate fauna on the Atlantic slope differs from the continental slopes of the mountain range; *Megalobulimus grandis* is believed to be restricted to the Atlantic side, but more research is needed on this species' distribution and ecology. Once more data is available, the conservation status of this species can be reassessed.

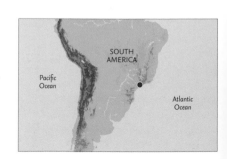

Brazilian Merganser

NOT EVALUATED	DATA DEFICIENT	LEAST CONCERN	NEAR THREATENED	VULNERABLE	ENDANGERED	CRITICALLY ENDANGERED	EXTINCT IN THE WILD	EXTINCT
NE	DD	LC	NT	VU	EN	CR	EW	EX

© ADRIANO GAMBARINI (WWW.GAMBARINI.COM.BR)

The **Brazilian Merganser**, Mergus octosetaceus, is listed as 'Critically Endangered' on the IUCN Red List of Threatened Species™. The highly fragmented global population of this riverine, fish-eating duck is believed to be fewer than 250 birds, found in the Cerrado biome of south-central Brazil. It has disappeared, or almost so, from Paraguay, Argentina and the Atlantic Forest of Brazil.

The contraction of this species' range is probably caused by a deterioration in river water quality. The Brazilian Merganser relies on clear water with healthy fish populations.

Deforestation and agricultural practices cause high rates of soil erosion, which creates sediment-laden watercourses in which this species cannot survive.

Three of the remaining sites for the species are now partially protected, and an Action Plan has been adopted by the Brazilian government. Birds are being studied in these sites, and around Canastra National Park there are also awareness-raising activities with local land users. Recovery of the species will need a transformation in land and water use practices, but such change would

substantially benefit human livelihoods, so there is great potential for positive action.

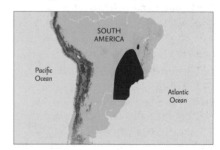

NOT EVALUATED	DATA DEFICIENT	LEAST CONCERN	NEAR THREATENED	‹ VULNERABLE ›	ENDANGERED	CRITICALLY ENDANGERED	EXTINCT IN THE WILD	EXTINCT
NE	DD	LC	NT	VU	EN	CR	EW	EX

The **Ceres Streamjack**, *Metacnemis angusta*, is classified as 'Vulnerable' on the IUCN Red List of Threatened Species™. It is a tiny damselfly that was first discovered in 1920, but was not seen again until it was rediscovered in 2003 in a pool adjacent to a river in the Cape Fold Mountains of South Africa. It is also commonly known as Ceres Featherlegs. Males are purple in colour and females are bright blue.

This species of damselfly is particularly at risk since it is only known to exist at one small pool, 50 m by 5 m in size. Despite intensive searches, the Ceres Streamjack has not been found elsewhere. This is partly due to the scarcity of its habitat within the region, and also because of invasive alien trees which threaten this damselfly by shading out its habitat.

Conservation measures have involved removal of invasive alien trees, but there is always a risk that they will regrow. An additional concern is that the only known locality is adjacent to a road where there is a risk from increased traffic emissions.

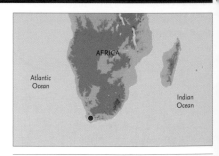

A species is Vulnerable when it faces a high risk of extinction in the wild, based on measurements of population size and/or geographic range and their trends in the past, present and/or future.

Cork Palm

NOT EVALUATED	DATA DEFICIENT	LEAST CONCERN	NEAR THREATENED	VULNERABLE	ENDANGERED	CRITICALLY ENDANGERED	EXTINCT IN THE WILD	EXTINCT
NE	DD	LC	NT	VU	EN	CR	EW	EX

The **Cork Palm**, Microcycas calocoma, is listed as 'Critically Endangered' on the IUCN Red List of Threatened Species™. It occurs in a small area of western Cuba in the province of Pinar del Río. In spite of its name (Microcycas means small cycad), this species has a trunk that can reach 15 m in height and is one of the tallest cycads in the world.

The Cork Palm grows in lowlands and hills with deep, stony-sandy soils, and in limestone cliffs with shallow soils and exposed rocks. The main threats to this species are habitat destruction and low levels of natural reproduction. Most of the remaining populations inhabiting limestone cliffs are located within two protected areas.

The Cork Palm is considered a 'Jewel of the Cuban Flora'. In 1989, it was declared a National Natural Monument and is protected under the Cuban Law of National and Local Monuments and CITES. A conservation plan aimed at increasing natural reproduction and recruitment is needed. Successful propagation means that the plant is quite well represented in botanical garden collections.

© J. LAZCANO LARA

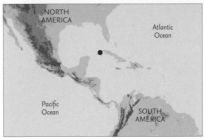

A species is **Critically Endangered** when it faces an extremely high risk of extinction in the wild, based on measurements of population size and/or geographic range and their trends in the past, present and/or future.

Mace Pagoda

NOT EVALUATED	DATA DEFICIENT	LEAST CONCERN	NEAR THREATENED	VULNERABLE	ENDANGERED	CRITICALLY ENDANGERED	EXTINCT IN THE WILD	EXTINCT
NE	DD	LC	NT	VU	EN	CR	EW	EX

The **Mace Pagoda**, Mimetes stokoei, is listed as 'Critically Endangered' on the Interim Red Data List of South African Plant Taxa. A member of the protea family, a group of plants characteristic of South Africa's fynbos shrublands, it has a life cycle adapted to fire.

The Mace Pagoda was previously known from a single population in the mountains bordering False Bay in the Western Cape. However, in 1965 an experimental plot was unknowingly established on top of the only existing population (the site was cleared and burned to make way for the new protea orchard); when no more plants were seen after 27 years and two fires, the species was presumed to be extinct. Fortunately, the Mace Pagoda's seeds turned out to remain viable underground for more than 50 years, and in 1999 a runaway wildfire created the very specific conditions needed for germination, producing 24 new seedlings.

The Mace Pagoda is still extremely vulnerable due to its small population size, restricted distribution, and extreme population fluctuations in response to fire. Urgent conservation measures are likely to be needed if the species is to survive.

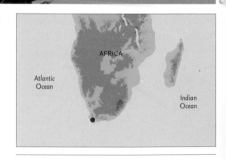

A species is Critically Endangered when it faces an extremely high risk of extinction in the wild, based on measurements of population size and/or geographic range and their trends in the past, present and/or future.

Floreana Mockingbird

RED LIST *Mimus trifasciatus*

NOT EVALUATED	DATA DEFICIENT	LEAST CONCERN	NEAR THREATENED	VULNERABLE	ENDANGERED	CRITICALLY ENDANGERED	EXTINCT IN THE WILD	EXTINCT
NE	DD	LC	NT	VU	EN	CR	EW	EX

The **Floreana Mockingbird**, *Mimus trifasciatus*, is classified as 'Critically Endangered' on the IUCN Red List of Threatened Species™. It survives today only on two tiny islets off the coast of Floreana Island in the Galápagos. Current population estimates are 47 birds on the 9-hectare Champion Island and 400–500 on the 80-hectare Gardner-por-Floreana.

The Floreana Mockingbird has a special place in science because it inspired Charles Darwin, during the HMS Beagle voyage, to develop the Theory of Natural Selection. In 1835, Darwin recorded the mockingbird as being common on Floreana, but it must have declined rapidly as the last reported sighting there was in 1868. Human hunting, predation by invasive mammals and disappearance of the prickly pear cactus, a favoured place to nest and a source of food, are thought to be responsible for the decline of this species.

A new effort to restore Floreana back to its former glory has been launched by the Galápagos National Park, aiming to remove invasive species from the main island and reintroduce marooned endemic species, including the Floreana Mockingbird.

© Michael Dvorak

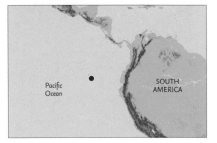

Pacific Ocean

SOUTH AMERICA

A species is **Critically Endangered** when it faces an extremely high risk of extinction in the wild, based on measurements of population size and/or geographic range and their trends in the past, present and/or future.

Demonic Poison Frog

NOT EVALUATED	DATA DEFICIENT	LEAST CONCERN	NEAR THREATENED	VULNERABLE	ENDANGERED	CRITICALLY ENDANGERED	EXTINCT IN THE WILD	EXTINCT
NE	DD	LC	NT	VU	EN	CR	EW	EX

The **Demonic Poison Frog** (or Yapacana's
Little Red Frog), Minyobates steyermarki,
is listed as 'Critically Endangered' on the
IUCN Red List of Threatened Species™.
It is a small but poisonous frog that
reaches lengths of only 14–16 mm. Its
range is probably less than 10 km², in
humid montane forest habitats on the
Cerro Yapacana, in southern Venezuela.

The Demonic Poison Frog is under
major threat from the destruction and
degradation of its habitat through opencast
gold-mining and its associated pollution
and fires, with additional habitat loss
resulting from wildfires. It is also under
threat from overcollection by the pet trade.

The Demonic Poison Frog is listed
on Appendix II of CITES, meaning
that trade in this species is regulated.
Furthermore, Cerro Yapacana is a
Venezuelan Natural Monument, which
may offer some protection to the species.

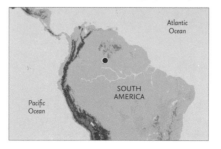

A species is **Critically Endangered** when it faces an
extremely high risk of extinction in the wild, based on
measurements of population size and/or geographic
range and their trends in the past, present and/or
future.

Giant Devilray

NOT EVALUATED	DATA DEFICIENT	LEAST CONCERN	NEAR THREATENED	VULNERABLE	⟨ ENDANGERED ⟩	CRITICALLY ENDANGERED	EXTINCT IN THE WILD	EXTINCT
NE	DD	LC	NT	VU	EN	CR	EW	EX

RED LIST

© Fabrizio Serena

The **Giant Devilray**, Mobula mobular, is listed as 'Endangered' on the IUCN Red List of Threatened Species™. This huge, plankton-feeding manta-like ray occurs in the Mediterranean and possibly also in the eastern Atlantic along the coast of northwest Africa, the Azores and the Canary Islands.

Although the Giant Devilray is not targeted specifically by fisheries, it is accidentally captured at threatening levels. Owing to high levels of by-catch and a very low reproductive capacity, this species is almost certainly declining. A decline in habitat quality, particularly in the Mediterranean, is also possibly impacting some populations.

The banning of driftnets and of trawling below 1,000 metres throughout the Mediterranean Sea in 2005 is likely to have reduced one of the most severe threats to the Giant Devilray. One of the major priorities now is to raise conservation awareness amongst fishermen in order to maximise the number of Giant Devilrays that are disentangled and released unharmed after accidental capture.

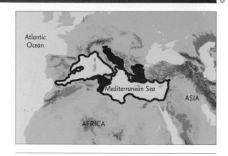

A species is Endangered when it faces a very high risk of extinction in the wild, based on measurements of population size and/or geographic range and their trends in the past, present and/or future.

Hawaiian Monk Seal

NOT EVALUATED	DATA DEFICIENT	LEAST CONCERN	NEAR THREATENED	VULNERABLE	ENDANGERED	CRITICALLY ENDANGERED	EXTINCT IN THE WILD	EXTINCT
NE	DD	LC	NT	VU	EN	CR	EW	EX

The **Hawaiian Monk Seal**, *Monachus schauinslandi*, is listed as 'Critically Endangered' on the IUCN Red List of Threatened Species™. It is the only true seal found year-round in tropical waters and, as its name suggests, it is found around the islands and atolls of Hawaii.

Historically, the Hawaiian Monk Seal was hunted for its meat, hide and oil. Additional pressure was experienced by the seals due to the presence of humans and dogs on islands and atolls that prevented them from using preferred habitats. Although now protected, the species is continuing to decline as a result of a lack of food resources, entanglement in marine debris and fishing gear, and shark predation. Rising sea level and acidification of reefs as a result of global warming could further impact this seal's future.

Additional efforts are needed to minimize interactions with humans in the inhabited main Hawaiian Islands, and consideration is being given to manipulation of the ecosystem of the northwestern Hawaiian Islands to make it more favourable to the seals (e.g. the removal of sharks).

A species is **Critically Endangered** when it faces an extremely high risk of extinction in the wild, based on measurements of population size and/or geographic range and their trends in the past, present and/or future.

NOT EVALUATED	DATA DEFICIENT	LEAST CONCERN	NEAR THREATENED	VULNERABLE	⟨ ENDANGERED ⟩	CRITICALLY ENDANGERED	EXTINCT IN THE WILD	EXTINCT
NE	DD	LC	NT	VU	EN	CR	EW	EX

© Tiit Maran

The **European Mink**, Mustela lutreola, is listed as 'Endangered' on the IUCN Red List of Threatened Species™. A century ago, the European Mink could be found throughout the European continent, but its population has severely declined and its range is now greatly reduced. Today, it is mostly known from Eastern Europe, with isolated populations in northern Spain and western France.

Historically, habitat loss and over-hunting are likely to have contributed to the European Mink's decline. In parts of Europe where hydroelectric developments and water pollution have increased significantly over the past few decades, habitat loss is still a serious threat. Pest-control trapping and accidental mortality through vehicle collisions also affect populations in some areas.

The European Mink is legally protected in all range states, and at least part of the population occurs within protected areas. In the early 1990s, an international conservation programme was set up by several partners. The programme's objective was to establish priorities for a conservation plan, by characterizing the bio-ecology of the species, analyzing causes for its decline and assessing the genetic variability of western populations.

A species is Endangered when it faces a very high risk of extinction in the wild, based on measurements of population size and/or geographic range and their trends in the past, present and/or future.

Indiana Bat

NOT EVALUATED	DATA DEFICIENT	LEAST CONCERN	NEAR THREATENED	VULNERABLE	⟨ ENDANGERED ⟩	CRITICALLY ENDANGERED	EXTINCT IN THE WILD	EXTINCT
NE	DD	LC	NT	VU	EN	CR	EW	EX

The **Indiana Bat**, Myotis sodalis, is listed as 'Endangered' on the IUCN Red List of Threatened Species™. It occurs in the Midwest and eastern United States, with northern populations migrating south for the winter.

This species has undergone a serious decline in recent years, with one of the main threats believed to be human disturbance at winter hibernation sites, which can cause direct mortality or cause the bats to rouse, depleting vital energy reserves. Indiana Bats have further been affected by the recently discovered White-nose Syndrome. At least 13,000 bats have died of this poorly understood disease.

A recovery plan is in place for the Indiana Bat, with conservation actions including management of hibernation sites and summer habitat, further research, and public education programmes. The most important hibernation sites are protected, but adequate protection of maternity roosts, prevention of human disturbance at roost sites, solving the White-nose Syndrome, and appropriate forest management, will be important for ensuring the long-term survival of this species.

© Merlin D Tuttle Bat Conservation International www.batcon.org

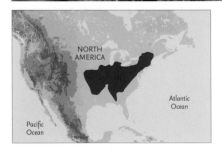

A species is **Endangered** when it faces a **very high risk** of extinction in the wild, based on measurements of population size and/or geographic range and their trends in the past, present and/or future.

NOT EVALUATED	DATA DEFICIENT	LEAST CONCERN	NEAR THREATENED	VULNERABLE	ENDANGERED	CRITICALLY ENDANGERED	EXTINCT IN THE WILD	EXTINCT
NE	DD	LC	NT	VU	EN	CR	EW	EX

© LOU JOST

Myriocolea irrorata is listed as 'Critically Endangered' on the IUCN Red List of Threatened Species™. This large, tropical liverwort (a group of species similar to mosses) was rediscovered in 2002, along the Topo River in the eastern foothills of the Ecuadorian Andes, at the original site where it had been collected almost 150 years earlier by the famous English botanist and explorer Richard Spruce.

The Topo River valley apparently creates optimal conditions for Myriocolea irrorata, conditions which are not found in other valleys in the region. Physical and chemical analyses of water suggest that this species may be extremely sensitive to environmental pollution. The construction of dams and dykes on the Topo River, proposed as part of a hydroelectric project, will alter water flow and quality which could be devastating for the remaining populations of this rare liverwort.

Since 2007 local conservationists have been fighting for conservation of the Topo River, and have won several lawsuits against the hydroelectric project based on the argument of the occurrence of Myriocolea irrorata along the river.

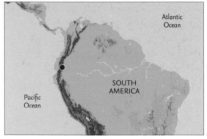

A species is Critically Endangered when it faces an extremely high risk of extinction in the wild, based on measurements of population size and/or geographic range and their trends in the past, present and/or future.

NOT EVALUATED	DATA DEFICIENT	LEAST CONCERN	NEAR THREATENED	⟨ VULNERABLE ⟩	ENDANGERED	CRITICALLY ENDANGERED	EXTINCT IN THE WILD	EXTINCT
NE	DD	LC	NT	VU	EN	CR	EW	EX

RED LIST

The **Giant Anteater**, *Myrmecophaga tridactyla*, is listed as 'Vulnerable' on the IUCN Red List of Threatened Species™. This unusual-looking mammal occupies a range extending from Honduras, south to Bolivia, Paraguay and Argentina, and feeds on ants and termites with the aid of its long, sticky tongue.

The dietary specificity, low reproductive rates and large body size of this species, along with habitat degradation, particularly in Central America, have proved to be significant factors in the Giant Anteater's decline. Individuals are also sometimes killed on roads, and where the species occurs in grassland habitats, it is at risk from both natural and human-caused fires. The Giant Anteater is also hunted for food, and in some areas it is hunted as a pest or to be kept as a pet.

This distinctive animal is protected across most of its range, and occurs in many protected areas. It is also listed on Appendix II of CITES, meaning that international trade in the species should be carefully regulated. Other recommended conservation measures for the Giant Anteater include improving fire management practices, particularly within the grassland areas it inhabits.

A species is Vulnerable when it faces a high risk of extinction in the wild, based on measurements of population size and/or geographic range and their trends in the past, present and/or future.

Eastern Sucker-footed Bat

Myzopoda aurita

NOT EVALUATED	DATA DEFICIENT	LEAST CONCERN		NEAR THREATENED	VULNERABLE	ENDANGERED	CRITICALLY ENDANGERED	EXTINCT IN THE WILD	EXTINCT
NE	DD	LC		NT	VU	EN	CR	EW	EX

© Merlin D. Tuttle, Bat Conservation International, www.batcon.org.

The **Eastern Sucker-footed Bat**, *Myzopoda aurita*, is listed as 'Least Concern' on the IUCN Red List of Threatened Species™. It has suckers on its wrists and hands which work by wet adhesion and give the bat its name. The Eastern Sucker-footed Bat is one of only two species of bat endemic to Madagascar. Until 2008, this species was listed as 'Vulnerable' on the IUCN Red List. However, following the discovery of a population in the southeast of the country in 2008, it was down-listed to 'Least Concern'.

Up to 50 individuals roost in the partially unfurled central leaf of the Traveller's Tree, *Ravenala madagascariensis*, but they move roosts every few days as the leaf unfurls. Unlike most mammals endemic to Madagascar, the Eastern Sucker-footed Bat is not threatened by deforestation, although harvesting of *Ravenala* leaves for building materials may cause the loss of roosts.

Only a few protected areas occur within the range of this species, including Parc National de Marojejy and Tampolo littoral forest. Additional studies are needed to increase understanding of local population densities and precise habitat requirements.

A species is **Least Concern** when it does not qualify for Critically Endangered, Endangered, Vulnerable or Near Threatened. Widespread and abundant species are included in this category.

AFRICA

Indian Ocean

NOT EVALUATED	DATA DEFICIENT	LEAST CONCERN	NEAR THREATENED	⟨ VULNERABLE ⟩	ENDANGERED	CRITICALLY ENDANGERED	EXTINCT IN THE WILD	EXTINCT
NE	DD	LC	NT	VU	EN	CR	EW	EX

Juliana's Golden Mole, *Neamblysomus julianae*, is listed as 'Vulnerable' on the IUCN Red List of Threatened Species™. This blind, subterranean mammal is endemic to South Africa, where it is only known from three isolated localities in the provinces of Gauteng, Limpopo and Mpumalanga.

The restricted savannah habitat of this species is under significant threat from human development, with intensive urbanization, mining operations, and agriculture being the greatest concerns. In particular, the subpopulation found on the slopes of the Bronberg Ridge in Tshwane (Pretoria), Gauteng, is at an extremely high risk of being lost in this area due to intensive urbanization and quarrying activities and, as a result, is listed separately on the IUCN Red List as 'Critically Endangered'.

Two of the subpopulations occur within protected areas; the Bronberg Ridge subpopulation is now protected by national and provincial legislation requiring detailed environmental impact assessments and mitigations before any developments may begin. The small area of remaining suitable habitat in the Bronberg, however, makes extinction risk for the subpopulation extremely high.

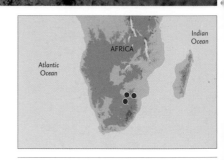

A species is **Vulnerable** when it faces a high risk of extinction in the wild, based on measurements of population size and/or geographic range and their trends in the past, present and/or future.

Kihansi Spray Toad

Nectophrynoides asperginis

NOT EVALUATED	DATA DEFICIENT	LEAST CONCERN	NEAR THREATENED	VULNERABLE	ENDANGERED	CRITICALLY ENDANGERED	EXTINCT IN THE WILD	EXTINCT
NE	DD	LC	NT	VU	EN	CR	EW	EX

The **Kihansi Spray Toad**, Nectophrynoides asperginis, is a dwarf toad, which rarely grows to be more than 2 cm long. It is now classified as 'Extinct in the Wild' on the IUCN Red List of Threatened Species™. This species was only known from a two-hectare range around the Kihansi Falls, in the Kihansi Gorge, eastern Tanzania.

The serious decline and extinction of this species appears related to habitat loss and the devastating amphibian fungal disease, chytridiomycosis. In 2000, a dam was constructed on the Kihansi River, which cut off 90 percent of the original water flow to the gorge. The volume of spray was significantly reduced, which also affected the vegetational composition of the Kihansi Spray Toad's habitat.

A captive breeding programme is ongoing in Toledo and New York Bronx Zoos. The programme also involves capacity building, in which Tanzanian experts are trained on amphibian husbandry techniques. These trained experts will then return to help build the breeding facility in Tanzania in preparation for Kihansi Spray Toad rehabilitation to their native wetland habitat.

© Timothy Herman

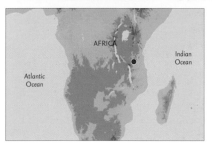

A species is Extinct in the Wild when it is known only to survive in cultivation or in captivity. A species is presumed Extinct in the Wild when exhaustive surveys in known and/or expected habitat, at appropriate times throughout its historic range, have failed to record an individual.

Cherry-throated Tanager

NOT EVALUATED	DATA DEFICIENT	LEAST CONCERN	NEAR THREATENED	VULNERABLE	ENDANGERED	CRITICALLY ENDANGERED	EXTINCT IN THE WILD	EXTINCT
NE	DD	LC	NT	VU	EN	CR	EW	EX

The **Cherry-throated Tanager**, *Nemosia rourei*, is listed as 'Critically Endangered' on the IUCN Red List of Threatened Species™. Previously feared to be extinct, this Atlantic Forest endemic was rediscovered in Espírito Santo State, Brazil, in 1998, a lapse of 47 years after the previous sighting.

Centuries of forest clearance for cattle ranches, plantations and timber have drastically reduced the indigenous Atlantic Forest of Brazil. Today, the Cherry-throated Tanager survives in isolated forest fragments at elevations greater than 850 metres above sea level, and what little remains of its habitat is under further threat from agricultural and urban encroachment and timber extraction.

With potentially fewer than 50 individuals remaining, the Cherry-throated Tanager is in drastic need of major conservation measures. Surveys have been conducted in recent years, the results of which demonstrate how rare the species is, along with a study of its ecology. Several privately owned forest tracts now need to be established as reserves in order to safeguard the long-term future of this enigmatic and elusive bird.

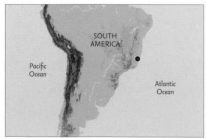

Santiago Galapagos Mouse

Nesoryzomys swarthi

NOT EVALUATED	DATA DEFICIENT	LEAST CONCERN	NEAR THREATENED	‹ VULNERABLE ›	ENDANGERED	CRITICALLY ENDANGERED	EXTINCT IN THE WILD	EXTINCT
NE	DD	LC	NT	VU	EN	CR	EW	EX

© Robert Dowler

The **Santiago Galapagos Mouse**, *Nesoryzomys swarthi*, is listed as 'Vulnerable' on the IUCN Red List of Threatened Species™. Believed extinct until its rediscovery in 1997, this species occurs only in a small area of the island of Santiago, in the Galapagos.

The endemic 'rice rats' have lost more species than any other group of vertebrates in the Galapagos, largely due to the introduction of Black Rats, mice and feral cats to the islands. Although it currently coexists with the Black Rat, the Santiago Galapagos Mouse is thought to owe its survival to areas of Opuntia cacti,

which act as refuges from competition, and to its greater ability to withstand drought. Unfortunately, climate change may allow Black Rat populations to increase dramatically, while at the same time decimating vital Opuntia habitat.

The most vital conservation measure for the Santiago Galapagos Mouse will be the eradication or control of introduced species. Projects are also underway to monitor and study the species, and an action plan has been proposed, together with the possibility of establishing a captive population.

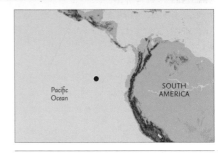

A species is Vulnerable when it faces a high risk of extinction in the wild, based on measurements of population size and/or geographic range and their trends in the past, present and/or future.

Luristan Newt

Neurergus kaiseri

NOT EVALUATED	DATA DEFICIENT	LEAST CONCERN	NEAR THREATENED	VULNERABLE	ENDANGERED	CRITICALLY ENDANGERED	EXTINCT IN THE WILD	EXTINCT
NE	DD	LC	NT	VU	EN	CR	EW	EX

The **Luristan Newt**, *Neurergus kaiseri*, is listed as 'Critically Endangered' on the IUCN Red List of Threatened Species™. It is a small species, reaching only 13 cm in length, and is found in the Luristan Province of Iran. It is thought that the striking mosaic of black and white patches on its fiery orange dorsal stripe serve to warn potential predators of its toxicity.

The Luristan Newt is estimated to number fewer than 1,000 mature individuals. It is threatened by habitat loss, recent severe droughts and the damming of the few inhabited streams. The greatest current concern for this species, however, is the increasing collection of wild individuals for the international pet trade.

This species is protected by Iranian national legislation, but immediate action is needed to prevent the illegal export of this attractive newt. Captive breeding programmes are being considered as a means of bolstering population numbers.

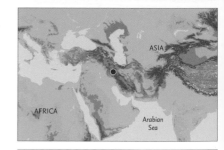

*A species is **Critically Endangered** when it faces an extremely high risk of extinction in the wild, based on measurements of population size and/or geographic range and their trends in the past, present and/or future.*

Bristlecone Hemlock

Nothotsuga longibracteata

NOT EVALUATED	DATA DEFICIENT	LEAST CONCERN	NEAR THREATENED	VULNERABLE	⟨ ENDANGERED ⟩	CRITICALLY ENDANGERED	EXTINCT IN THE WILD	EXTINCT
NE	DD	LC	NT	VU	EN	CR	EW	EX

© Y LIU, INSTITUTE OF BOTANY OF CHINESE ACADEMY OF SCIENCES, BEIJING

Bristlecone Hemlock, *Nothotsuga longibracteata*, is listed as 'Endangered' on the IUCN Red List of Threatened Species™. This tree grows up to 30 metres in height and is found in China in areas with mountains at elevations of between 300 and 2,300 metres above sea level.

This species is very rare despite its relatively wide distribution. Large scale logging has depleted the number of trees to an undetermined extent, and substantial parts of forest containing this species have disappeared especially at lower elevations.

In 1982, Bristlecone Hemlock was found in the far northwest of Yunnan by a Chinese team; this identification was verified in October 2000. This disjunctive occurrence at high altitude could indicate that its actual distribution remains unknown.

In China, Bristlecone Hemlock is considered to be a desirable forest tree suitable for afforestation. However, for this species to be conserved, its use as a timber tree must be limited, particularly as it is not in general cultivation outside China, and is rare in botanical collections.

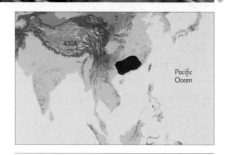

A species is **Endangered** when it faces a *very high risk* of extinction in the wild, based on measurements of population size and/or geographic range and their trends in the past, present and/or future.

Maathai's Longleg

Notogomphus maathaiae

NOT EVALUATED	DATA DEFICIENT	LEAST CONCERN	NEAR THREATENED	VULNERABLE	‹ ENDANGERED ›	CRITICALLY ENDANGERED	EXTINCT IN THE WILD	EXTINCT
NE	DD	LC	NT	VU	EN	CR	EW	EX

Maathai's Longleg, *Notogomphus maathaiae*, is classified as 'Endangered' on the IUCN Red List of Threatened Species™. This dragonfly was named in honour of Wangari Maathai, the Nobel Peace laureate for 2004 and the first African woman to be awarded this prize. Wangari was recognised for her efforts to advocate sustainable human development by focusing on the protection of Africa's last remaining and fast-shrinking forests. This species can only be found in the Kenyan highlands, where Maathai was born and worked.

Maathai's Longleg is only known from clear streams in three high-mountain forests. Such habitats have been widely destroyed in recent decades and, as deforestation continues in the densely populated highlands, this rare dragonfly may become Critically Endangered. It therefore serves as an indicator of healthy streams and as a flagship species for urgent watershed protection.

Conserving riverside forests not only helps this dragonfly, but also the people of the foothills, by guaranteeing stable soils and a clean and dependable water source. To this end, dragonflies are dubbed 'guardians of the watershed'.

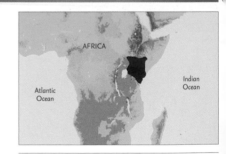

A species is **Endangered** when it faces a very high risk of extinction in the wild, based on measurements of population size and/or geographic range and their trends in the past, present and/or future.

RED LIST

NOT EVALUATED	DATA DEFICIENT	LEAST CONCERN	NEAR THREATENED	VULNERABLE	ENDANGERED	CRITICALLY ENDANGERED	‹ EXTINCT IN THE WILD ›	EXTINCT
NE	DD	LC	NT	VU	EN	CR	EW	EX

Nymphaea thermarum is listed as 'Extinct in the Wild' on the IUCN Red List of Threatened Species™. It is the world's smallest water lily with leaves of only 1 cm in size, and tiny white flowers with bright yellow stamens. This wetland plant was only known from one site of damp mud formed by the overflow of a hot spring in Mashyuza, Rwanda.

Nymphaea thermarum became extinct in 2008 when its habitat was destroyed by farmers who used the aquifer that fed the hot spring for agriculture. Fortunately, numerous seeds were collected before the species disappeared. Botanists were unable to germinate any seeds until senior horticulturist Carlos Magdalena, from Kew Royal Botanic Gardens in London, found out the secret in 2009. While water lily species typically germinate deep under water, this particular species needed to be close to the surface.

The species is now propagated in Kew Gardens and is a potential candidate for a reintroduction programme.

© Andrew McRobb RBG, Kew

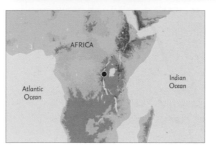

A species is Extinct in the Wild when it is known only to survive in cultivation or in captivity. A species is presumed Extinct in the Wild when exhaustive surveys in known and/or expected habitat, at appropriate times throughout its historic range, have failed to record an individual.

NOT EVALUATED	DATA DEFICIENT	LEAST CONCERN	NEAR THREATENED	VULNERABLE	⟨ ENDANGERED ⟩	CRITICALLY ENDANGERED	EXTINCT IN THE WILD	EXTINCT
NE	DD	LC	NT	VU	EN	CR	EW	EX

The **Ili Pika**, Ochotona iliensis, is listed as 'Endangered' on the IUCN Red List of Threatened Species™. This small lagomorph is endemic to China, where it has a highly fragmented distribution within a restricted geographical range in the Tien Shan Mountains.

Over the past ten years, the Ili Pika has suffered a dramatic decline, both in terms of its population size and the extent of its range. The exact cause of this decline is not known, but it is speculated that climate change and an increase in grazing pressure are having a negative impact on the species. Furthermore, low population densities and reproductive rates, coupled with a relatively limited ability to disperse, are impeding the ability of the species to recover.

There are no known conservation measures in place for the Ili Pika. Further research needs to be conducted, and the implementation of a recovery plan is vital if this species is to be prevented from becoming extinct.

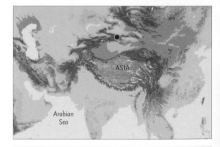

A species is **Endangered** when it faces a *very high risk* of extinction in the wild, based on measurements of population size and/or geographic range and their trends in the past, present and/or future.

Okapi

Okapia johnstoni

NOT EVALUATED	DATA DEFICIENT	LEAST CONCERN	NEAR THREATENED	VULNERABLE	ENDANGERED	CRITICALLY ENDANGERED	EXTINCT IN THE WILD	EXTINCT
NE	DD	LC	NT	VU	EN	CR	EW	EX

© Bob Jenkins

The **Okapi**, Okapia johnstoni, is listed as 'Near Threatened' on the IUCN Red List of Threatened Species™. Although rather horse-like in appearance, this intriguing-looking mammal is, in fact, a relative of the Giraffe, *Giraffa camelopardalis*. It inhabits forests of the Democratic Republic of Congo, and was also previously known from Uganda, where it is now thought to be extinct.

The main threat to the Okapi is habitat loss in the form of logging and human settlement. Hunting for meat and skins is also a problem, and this species has declined rapidly in areas where cable snares are persistently used.

The Okapi is officially protected under Congolese law, and is also a national symbol of the Democratic Republic of Congo. Significant populations occur in the Okapi Faunal Reserve and Maiko National Park, and the effective protection of these two areas has been identified as the most important measure for the Okapi's long-term survival.

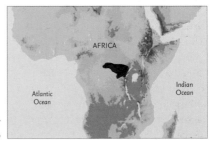

AFRICA

Atlantic Ocean

Indian Ocean

NOT EVALUATED	DATA DEFICIENT	LEAST CONCERN	NEAR THREATENED	VULNERABLE	ENDANGERED	CRITICALLY ENDANGERED	EXTINCT IN THE WILD	EXTINCT
NE	DD	LC	NT	VU	EN	CR	EW	EX

Opuntia chaffeyi is listed as 'Critically Endangered' on the IUCN Red List of Threatened Species™. This small branching cactus is restricted to a small area in the state of Zacatecas, Mexico, where it is only known from three localities, each comprising no more than five plants.

With a population potentially numbering no more than 15 mature plants, Opuntia chaffeyi is extremely close to becoming extinct. Habitat destruction due to cattle overpasturing is the major threat, while parts of this species are also harvested for medicinal purposes, as an anti-inflammatory. There is also concern that the area could be impacted upon by agricultural expansion if boreholes are sunk to obtain water for irrigation.

Although Opuntia chaffeyi is listed on Appendix II of CITES, which prohibits international trade without a permit, further conservation measures are clearly necessary. Probably the highest priority is the creation of protected areas, which would also benefit other threatened species with which this rare cactus grows.

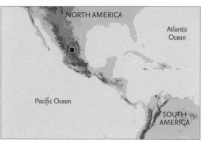

A species is **Critically Endangered** when it faces an **extremely high risk** of extinction in the wild, based on measurements of population size and/or geographic range and their trends in the past, present and/or future.

RED LIST

NOT EVALUATED	DATA DEFICIENT	LEAST CONCERN	NEAR THREATENED	‹ VULNERABLE ›	ENDANGERED	CRITICALLY ENDANGERED	EXTINCT IN THE WILD	EXTINCT
NE	DD	LC	NT	VU	EN	CR	EW	EX

© Steve Woodhall

The **Karkloof Blue**, *Orachrysops ariadne*, is classified as 'Vulnerable' on the IUCN Red List of Threatened Species™. It is confined to a very special type of habitat, wood's indigo, in KwaZulu-Natal, South Africa. While it lays its eggs on the host plant, the larva grows up in association with the Carpenter ants which live alongside it. Its four populations are very small, with only a few dozen individuals at each locality.

The main threat to the Karkloof Blue is the conversion of land to commercial timber plantations, with a simultaneous spread of these alien trees into natural areas. Inappropriate burning regimes within this species' habitat, and the impacts of domestic cattle, are also potential threats to their very small populations.

The four localities are now in protected areas, where the butterfly is protected and the habitat is managed appropriately. At one of the localities, Nkandla Forest Reserve, the local Chube community is overseeing the care of this flagship species. The Karkloof Blue is also represented on a logo for the local tourist route, the Midlands Meander.

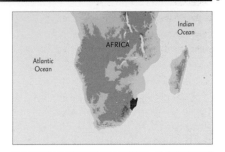

A species is Vulnerable when it faces a high risk of extinction in the wild, based on measurements of population size and/or geographic range and their trends in the past, present and/or future.

Irrawaddy Dolphin

Orcaella brevirostris

NOT EVALUATED	DATA DEFICIENT	LEAST CONCERN	NEAR THREATENED	⟨ VULNERABLE ⟩	ENDANGERED	CRITICALLY ENDANGERED	EXTINCT IN THE WILD	EXTINCT
NE	DD	LC	NT	VU	EN	CR	EW	EX

The **Irrawaddy Dolphin**, *Orcaella brevirostris*, is listed as 'Vulnerable' on the IUCN Red List of Threatened Species™. A distinctive dolphin with a rounded head and no beak, this species is patchily distributed in shallow, coastal waters of the Indo-Pacific, and also occurs in three large river systems – the Ayeyarwady in Myanmar, the Mekong in Cambodia, and the Mahakam in Borneo, Indonesia.

The main threat to the Irrawaddy Dolphin is entanglement in gill nets. Habitat loss and degradation from pollution, dam construction, sedimentation, and vessel traffic are additional concerns. These dolphins are revered in many parts of Asia. Five subpopulations, including all three freshwater populations, have suffered dramatic declines in range and numbers and are considered Critically Endangered.

The Irrawaddy Dolphin is legally protected from deliberate capture in most of its range, and specially protected areas have been designated in several areas. Restrictions on the use of gill nets are necessary to reduce entanglement mortality. The use of non-entangling and more selective gear is to be encouraged.

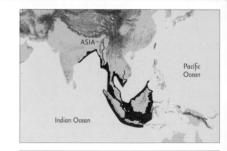

A species is Vulnerable when it faces a high risk of extinction in the wild, based on measurements of population size and/or geographic range and their trends in the past, present and/or future.

NOT EVALUATED	DATA DEFICIENT	LEAST CONCERN	NEAR THREATENED	VULNERABLE	‹ ENDANGERED ›	CRITICALLY ENDANGERED	EXTINCT IN THE WILD	EXTINCT
NE	DD	LC	NT	VU	EN	CR	EW	EX

© GEORGE TURNER

Oreochromis karongae is listed as 'Endangered' on the IUCN Red List of Threatened Species™. This fish is one of three 'chambo' species endemic to Lake Malawi, and can be seen in loose shoals in many areas of the lake. Like other Oreochromis they are maternal mouthbrooders, with the mother protecting the eggs in her mouth until they hatch, and often the juvenile fry continue to use her mouth for shelter when danger is present.

The chambo are the most valuable food fishes in Malawi, and are very important for local subsistence as well as commerce. Populations of this species, however, collapsed in the 1990s as a result of overfishing, with as much as a 70% decline within 10 years.

This species is, however, very easy to maintain in captivity. Although formal management of fisheries in the lake has been in place since the 1930s, over-harvesting is still a problem. Greater enforcement of harvest levels is vital in order to minimize further losses of this species and other fishes in Lake Malawi.

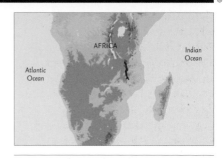

A species is **Endangered** when it faces a very high risk of extinction in the wild, based on measurements of population size and/or geographic range and their trends in the past, present and/or future.

Scimitar-horned Oryx

Oryx dammah

NOT EVALUATED	DATA DEFICIENT	LEAST CONCERN	NEAR THREATENED	VULNERABLE	ENDANGERED	CRITICALLY ENDANGERED	EXTINCT IN THE WILD	EXTINCT
NE	DD	LC	NT	VU	EN	CR	EW	EX

The **Scimitar-horned Oryx**, *Oryx dammah*, is listed as 'Extinct in the Wild' on the IUCN Red List of Threatened Species™. Named for its magnificently curved horns, it was once widespread all around the Sahara desert, on the northern fringe, and in the Sahel zone from the Atlantic Ocean to the Red Sea, but is believed to have been lost from the wild in 1999.

The original decline of this species is thought to have been caused by major climatic changes that led to the expansion of the Sahara desert. However, its ultimate demise in the wild has been attributed to overhunting for its meat, hide and horns.

The only remaining Scimitar-horned Oryx occur in a number of captive populations, very few of which occur in its former range. Some are in zoos, and large numbers are being kept in private collections in the Arabian Peninsula and in private game ranches in the USA, South Africa and other countries. A captive breeding programme is in place and, as part of a reintroduction project, individuals have been released into fenced protected areas in Tunisia, Morocco and Senegal, with further reintroductions planned.

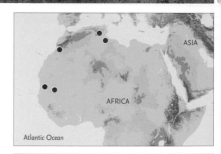

A species is Extinct in the Wild when it is known only to survive in cultivation or in captivity. A species is presumed Extinct in the Wild when exhaustive surveys in known and/or expected habitat, at appropriate times throughout its historic range, have failed to record an individual.

NOT EVALUATED	DATA DEFICIENT	LEAST CONCERN	NEAR THREATENED	VULNERABLE	⟨ ENDANGERED ⟩	CRITICALLY ENDANGERED	EXTINCT IN THE WILD	EXTINCT
NE	DD	LC	NT	VU	EN	CR	EW	EX

© SOLO HERY J.V. RAPANARIVO

Pachypodium windsorii has not yet been evaluated for the IUCN Red List of Threatened Species™, however it has provisionally been assessed as 'Endangered'. It is a succulent plant found in the extreme north of Madagascar. For a long time, it was only known to occur in the Windsor Castle massif but, in 2005, two new subpopulations were discovered at the Beantely massif and Amboaizamikono massif.

This species grows either in forest or wooded grassland, but always on calcareous rock outcrops or cliffs. The area has been severely impacted by forest logging and fire, which have considerably reduced the primary vegetation to small patches of secondary dry forest amidst a landscape dominated by grassland or wooded grassland. Exploitation of these fragments of forest for timber and fuel continues.

The only conservation measure currently being undertaken for Pachypodium windsorii is the ex situ (i.e. conservation outside their natural habitat) transplantation of seedlings to the Tsimbazaza Botanical Garden, Antananarivo, Madagascar.

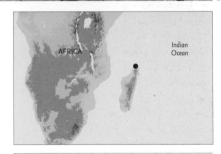

A species is Endangered when it faces a very high risk of extinction in the wild, based on measurements of population size and/or geographic range and their trends in the past, present and/or future.

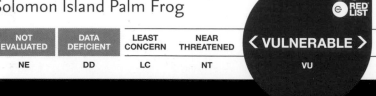

NOT EVALUATED	DATA DEFICIENT	LEAST CONCERN	NEAR THREATENED	‹ VULNERABLE ›	ENDANGERED	CRITICALLY ENDANGERED	EXTINCT IN THE WILD	EXTINCT
NE	DD	LC	NT	VU	EN	CR	EW	EX

The **Solomon Island Palm Frog**, *Palmatorappia solomonis*, is listed as 'Vulnerable' on the IUCN Red List of Threatened Species™. This species inhabits tropical rainforest on several large islands which straddle the border between Papua New Guinea and the Solomon Islands in the Western Pacific.

In just the last three decades, the clearance and degradation of forest has left only 25 percent of the region's lowland forest in a pristine condition,

and it is thought that logging might be impacting lowland populations of the Solomon Island Palm Frog.

Given the current rate of forest loss within the Solomon Islands, there is an urgent need to improve habitat protection at sites where the Solomon Island Palm Frog occurs. At present, it is not known to inhabit any protected areas within its range.

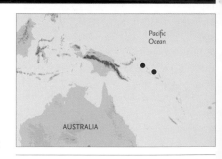

A species is **Vulnerable** when it faces a high risk of extinction in the wild, based on measurements of population size and/or geographic range and their trends in the past, present and/or future.

Le Pouce Mountain Screwpine

NOT EVALUATED	DATA DEFICIENT	LEAST CONCERN	NEAR THREATENED	VULNERABLE	ENDANGERED	CRITICALLY ENDANGERED	EXTINCT IN THE WILD	EXTINCT
NE	DD	LC	NT	VU	EN	CR	EW	EX

© VINCENT FLORENS

The **Le Pouce Mountain Screwpine**, *Pandanus pseudomontanus*, is not currently listed on the IUCN Red List of Threatened Species™, however its conservation status has been potentially assessed as 'Critically Endangered'. It is unique to the island of Mauritius and, since it is known only from two male plants in the Le Pouce Mountain Nature Reserve, it faces an extremely high risk of extinction.

The Screwpine's habitat is at the edge of a small patch of wet forest surviving on the mountain's flank. This forest was largely destroyed in the 19th century (Charles Darwin made reference to this deforestation in his diary during his stop in Mauritius). Rats, an invasive alien in Mauritius, are known to destroy the seeds of certain Screwpine species and may have contributed to the decline of the plant.

With no female plant known, the only chance for the survival of this species could be through cloning and sex change attempts (which have been successful with another unique Mauritian plant). This will require close collaboration with adequately equipped and staffed foreign institutions, and should be implemented as a matter of urgency.

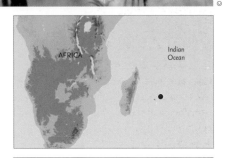

A species is **Critically Endangered** when it faces an extremely high risk of extinction in the wild, based on measurements of population size and/or geographic range and their trends in the past, present and/or future.

Mekong Giant Catfish

·NOT EVALUATED	DATA DEFICIENT	LEAST CONCERN	NEAR THREATENED	VULNERABLE	ENDANGERED	CRITICALLY ENDANGERED	EXTINCT IN THE WILD	EXTINCT
NE	DD	LC	NT	VU	EN	CR	EW	EX

The **Mekong Giant Catfish**, *Pangasianodon gigas*, is classified as 'Critically Endangered' on the IUCN Red List of Threatened Species™. It is the world's largest freshwater fish and is found only in the parts of the Mekong River basin that run through Cambodia, Laos, Thailand, Vietnam and possibly Burma and China.

The Mekong Giant Catfish has been subject to overfishing for many years. As a result of damming and clearance of the flooded forest near the Tonle Sap Lake, its habitat has been severely disrupted effecting its migration, spawning, eating and breeding habits.

Legislation restricting the hunting of Mekong Giant Catfish exists but is rarely enforced. Artificially spawned individuals have been released into the River Mekong since 1985, and captive breeding (reliant on wild-caught brood stock) has been taking place since 2001. The potentially highly significant impact of dams needs urgent assessment, as recent studies suggest that all large migratory catfish will be eliminated from the river system if two or more mainstream dams are constructed without effective adaptation measures.

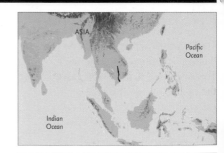

A species is **Critically Endangered** when it faces an extremely high risk of extinction in the wild, based on measurements of population size and/or geographic range and their trends in the past, present and/or future.

Tiger

Panthera tigris

RED LIST

NOT EVALUATED	DATA DEFICIENT	LEAST CONCERN	NEAR THREATENED	VULNERABLE	‹ ENDANGERED ›	CRITICALLY ENDANGERED	EXTINCT IN THE WILD	EXTINCT
NE	DD	LC	NT	VU	EN	CR	EW	EX

© ALEX S.LWA

The **Tiger**, *Panthera tigris*, is listed as 'Endangered' on the IUCN Red List of Threatened Species™. The largest of all cats, the Tiger once occurred throughout central, eastern and southern Asia, but currently survives only in scattered populations.

The Caspian, Javan and Bali Tigers are already extinct, and of the remaining six subspecies, the South China Tiger has not been observed for many years. With approximately 1,400 individuals, India still has the largest national population; however, globally, no more than about 3,200 Tigers roam free in their natural habitat. Poaching and illegal killing are the major threats to the survival of the remaining populations, but habitat loss and overhunting of Tigers and their natural prey species have caused a reduction in distribution, which is now only seven percent of the historic range.

The key to this species' survival is the immediate protection of the remaining populations and, in the long-term, the maintenance or recovery of large tracts of habitat and corridors, together with the sustainable management of prey populations. This will only be possible through mitigation of the conflicts between local people and Tiger conservation.

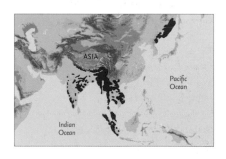

Imperiled Katydid

NOT EVALUATED	DATA DEFICIENT	LEAST CONCERN	NEAR THREATENED	VULNERABLE	ENDANGERED	CRITICALLY ENDANGERED	EXTINCT IN THE WILD	EXTINCT
NE	DD	LC	NT	VU	EN	CR	EW	EX

The **Imperiled Katydid**, Paracilacris periclitatus, is currently not officially listed on the IUCN Red List of Threatened Species™. However, its conservation status has been assessed as potentially 'Critically Endangered'. It was recently discovered in an isolated patch of indigenous forest surrounded by exotic timber plantations in KwaZulu-Natal Province, South Africa. To date, no other populations of this species have been found.

South Africa's grassland biome is sustained by frequent natural fires. In the absence of fire, grassland quickly turns into indigenous forest, so the landscape contains small patches of forest surrounded by a grassland matrix. However, the grassland matrix is rapidly being replaced by commercial timber plantations of exotic tree species from Europe and Australia, which grow rapidly in the South African climate.

Plantations are inhospitable to the Imperiled Katydid's probable food source, a common grass species found in transition zones between grassland and indigenous forest. Due to its reduced wings and small body size, the Imperiled Katydid has limited dispersal capabilities and may not easily colonize new areas should its present habitat be destroyed.

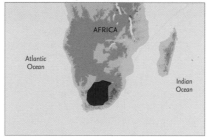

A species is **Critically Endangered** when it faces an extremely high risk of extinction in the wild, based on measurements of population size and/or geographic range and their trends in the past, present and/or future.

Marbled Cat

Pardofelis marmorata

The **Marbled Cat**, *Pardofelis marmorata*, is listed as 'Vulnerable' on the IUCN Red List of Threatened Species™. This enigmatic predator is found from northern India and Nepal, through south-eastern Asia to Borneo and Sumatra.

The Marbled Cat is primarily diurnal, although it is also active at night. It is thought to be rare throughout its range, although infrequent encounters may be attributable to its reclusive nature and remote forest habitat. In particular, it is the Marbled Cat's affinity for arboreality (spending most or all of its time in trees) that makes encounters so rare when compared with, for example, the Clouded Leopard or Leopard Cat. The major threat to this species is believed to be the widespread destruction of forest habitat throughout Southeast Asia, which also reduces its prey base.

The Marbled Cat occurs in a number of protected areas and hunting is prohibited or regulated throughout much of its range. However, with so little known about this species, further investigation into its status in the wild, and the degree to which it can tolerate loss and disturbance of its forest habitat, is urgently needed.

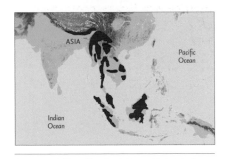

A species is Vulnerable when it faces a high risk of extinction in the wild, based on measurements of population size and/or geographic range and their trends in the past, present and/or future.

South African Geranium

Pelargonium sidoides

NOT EVALUATED	DATA DEFICIENT	LEAST CONCERN		NEAR THREATENED	VULNERABLE	ENDANGERED	CRITICALLY ENDANGERED	EXTINCT IN THE WILD	EXTINCT
NE	DD	LC		NT	VU	EN	CR	EW	EX

The **South African Geranium**, *Pelargonium sidoides*, locally called Umckaloabo, has a provisional assessment of 'Least Concern' on the IUCN Red List of Threatened Species™. It is a member of the Geranium family and is an aromatic perennial herb endemic to South Africa and Lesotho, where it is widely distributed in open grasslands.

Pelargonium species have long been used in local traditional remedies for colic, dysentery, and other abdominal ailments. Recently, the South African Geranium has come under severe harvest pressure in order to satisfy the growing international market for the plant's underground branches. These are used in commercially produced remedies to treat bronchitis and other respiratory tract infections. This species is also threatened by overgrazing and poor management of rangelands.

There is no current legal protection, monitoring, or management system in place for the South African Geranium. However, the two range state governments have collaborated with TRAFFIC East/Southern Africa, and an industry partner, to implement the sustainable commercial collection of this species in Lesotho.

© DAVID NEWTON

A species is **Least Concern** when it does not qualify for Critically Endangered, Endangered, Vulnerable or Near Threatened. Widespread and abundant species are included in this category.

AFRICA

Atlantic Ocean

Indian Ocean

Dalmatian Pelican

Pelecanus crispus

NOT EVALUATED	DATA DEFICIENT	LEAST CONCERN	NEAR THREATENED	VULNERABLE	ENDANGERED	CRITICALLY ENDANGERED	EXTINCT IN THE WILD	EXTINCT
NE	DD	LC	NT	VU	EN	CR	EW	EX

© Giorgos Catsadorakis

The **Dalmatian Pelican**, *Pelecanus crispus*, is listed as 'Vulnerable' on the IUCN Red List of Threatened Species™, and has a decreasing population trend. Its worldwide population is estimated to be 10,000–20,000 individuals including 4,000–5,000 breeding pairs. It breeds in wetlands and has a wide but fragmented distribution from east and southeast Europe, to Mongolia and up to the coasts of China and Hong Kong.

Former declines of the Dalmatian Pelican were mainly caused by wetland drainage and persecution by fishermen, with continuing threats that include disturbance from tourists and fishermen, and destruction and alteration of wetland habitats. Other threats to this species include water pollution, collision with overhead power lines and overexploitation of fish stocks.

Main conservation measures should include surveys carried out on breeding and wintering grounds in west, central and east Asia, and especially all former Soviet Union countries, where the largest part of the global population is found. Sustainable management of wetlands is required across the range of the Dalmatian Pelican, but more specifically, establishment of non-intrusion zones around breeding colonies would solve most problems.

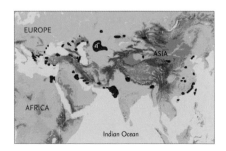

Amami Rabbit

Pentalagus furnessi

NOT EVALUATED	DATA DEFICIENT	LEAST CONCERN	NEAR THREATENED	VULNERABLE	‹ ENDANGERED ›	CRITICALLY ENDANGERED	EXTINCT IN THE WILD	EXTINCT
NE	DD	LC	NT	VU	EN	CR	EW	EX

The **Amami Rabbit**, *Pentalagus furnessi*, is listed as 'Endangered' on the IUCN Red List of Threatened Species™. Having evolved in isolation on two small Japanese islands lacking mammalian predators, this heavy-bodied species has a very distinctive appearance.

Owing to widespread habitat degradation brought about by logging and development, the Amami Rabbit population has undergone a significant decline, with only four fragmented subpopulations remaining. Since 1980, the amount of old growth forest on the islands of Amami and Tokuno has declined by an alarming 70 to 90 percent. In addition, predation by introduced mammalian predators, such as dogs, cats and mongooses, poses a further threat to the survival of this species.

The Amami Rabbit is classified as a Japanese National Monument, and as such receives protection from hunting and capture. However, habitat loss and invasive species are currently the major threats to its survival, and the protection of remaining forests and management of invasive species will be the key to securing the future of this unique rabbit.

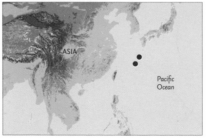

A species is Endangered when it faces a very high risk of extinction in the wild, based on measurements of population size and/or geographic range and their trends in the past, present and/or future.

Day Gecko

NOT EVALUATED	DATA DEFICIENT	LEAST CONCERN	NEAR THREATENED	VULNERABLE	ENDANGERED	CRITICALLY ENDANGERED	EXTINCT IN THE WILD	EXTINCT
NE	DD	LC	NT	VU	EN	CR	EW	EX

The **Day Gecko**, *Phelsuma antanosy*, is listed as 'Critically Endangered' on the IUCN Red List of Threatened Species™. This highly specialized gecko forms a close association with a single species of *Pandus* tree in which it lays its eggs. Consequently, the Day Gecko is only found in the coastal forests of the Tolagnaro region in southeastern Madagascar that support this tree species.

The Day Gecko has a highly fragmented distribution, and is thought to occupy an area of no more than nine square kilometres. Its remaining habitat is further threatened by tree felling for timber, fuel and conversion to agricultural land, and the selective harvesting of its host tree species. Furthermore, one subpopulation at Sainte Luce is expected to be lost as mining for ilmenite begins.

Protected areas have been created within the Day Gecko's range, and five forest fragments in Sainte Luce are managed within a community resource use agreement, which aims to sustain local livelihoods whilst conserving biodiversity. However, the successful management of these areas will very much depend on the involvement and support of the local communities.

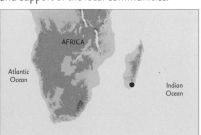

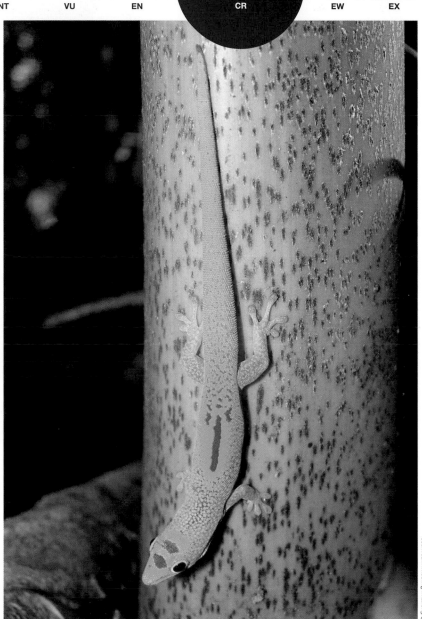

© Christian Randrianantoandro

Steiner's Shrub Frog

Philautus steineri

RED LIST

NOT EVALUATED	DATA DEFICIENT	LEAST CONCERN	NEAR THREATENED	VULNERABLE	‹ ENDANGERED ›	CRITICALLY ENDANGERED	EXTINCT IN THE WILD	EXTINCT
NE	DD	LC	NT	VU	EN	CR	EW	EX

Steiner's Shrub Frog, Philautus steineri, was first listed as 'Endangered' by the IUCN Red List of Threatened Species™ in 2006 and this classification has remained unchanged. Steiner's Shrub Frog is endemic to Sri Lanka and occupies an area smaller than 500 km² in the Knuckles Mountain Range.

The major threat to Steiner's Shrub Frog is the clearing of its forest habitat. Logging, as well as smallholder agricultural activities, have resulted in a declining population in recent years. The whole Knuckles Forest region was declared as a National Man and Biosphere Reserve by the Sri Lankan Government in April 2000, offering Steiner's Shrub Frog some habitat protection.

There is still, however, an urgent need for further protection of forest habitats in the Corbett's Gap region of Sri Lanka, and for continued monitoring of the species' population in order to prevent further reduction in the number of Steiner's Shrub Frogs.

This frog is named in honour of Achim Steiner, Executive Director of the UN Environment Programme.

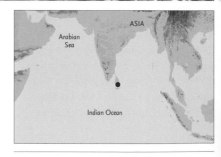

A species is Endangered when it faces a very high risk of extinction in the wild, based on measurements of population size and/or geographic range and their trends in the past, present and/or future.

Vaquita

Phocoena sinus

NOT EVALUATED	DATA DEFICIENT	LEAST CONCERN	NEAR THREATENED	VULNERABLE	ENDANGERED	CRITICALLY ENDANGERED	EXTINCT IN THE WILD	EXTINCT
NE	DD	LC	NT	VU	EN	CR	EW	EX

© Thomas A. Jefferson. Photo taken under permit (Oficio No. DR_488/08) from the Secretaría de Medio Ambiente y Recursos Naturales (SEMARNAT), within a natural protected area subject to special management and decreed as such by the Mexican Government.

The **Vaquita**, *Phocoena sinus*, is listed as 'Critically Endangered' on the IUCN Red List of Threatened Species™. This small porpoise is restricted to the upper Gulf of California, Mexico, and has the unfortunate distinction of being the world's most threatened cetacean species.

With a population estimated at only around 150, the Vaquita is undergoing a disastrous decline. The upper Gulf of California is intensively fished, and as a result the most serious threat to the Vaquita is entanglement in gill nets.

Since 2008, the Government of Mexico has established the Vaquita Refuge as a net-free fishing zone, and reduced the level of fishing effort in other areas through a combination of economic measures. Compensation for fishing permit holders who retire their permits or switch to other Vaquita-safe fishing methods, and improved enforcement to regulate illegal fishing have so far reduced the gill net fishing effort by about a third. However, given the perilous status of the Vaquita, stronger measures are needed to ensure that all gill nets are removed from the entire range of the species.

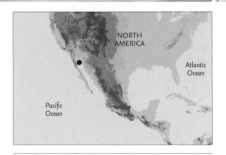

A species is Critically Endangered when it faces an extremely high risk of extinction in the wild, based on measurements of population size and/or geographic range and their trends in the past, present and/or future.

NOT EVALUATED	DATA DEFICIENT	LEAST CONCERN	NEAR THREATENED	VULNERABLE	ENDANGERED	CRITICALLY ENDANGERED	EXTINCT IN THE WILD	EXTINCT
NE	DD	LC	NT	VU	EN	CR	EW	EX

Phyllomedusa ayeaye is listed as 'Critically Endangered' on the IUCN Red List of Threatened Species™. This small frog is identified by its green body and pattern of red-orange blotches, encircled by black or purple, on its flanks and limbs. It is currently known only from small areas in the Brazilian states of Minas Gerais and São Paulo.

Phyllomedusa ayeaye is found in the transition zone between cerrado (tropical woodland savanna) and Atlantic semi-deciduous forest, laying its eggs on leaves above streams or pools so that the tadpoles, when hatched, fall into the water below. This species is under threat from habitat loss resulting from mining activity and fires, and is also impacted by pollution from mining and pesticides. Its restricted range is likely to make it particularly vulnerable to these threats.

While there are currently no specific conservation measures targetting this colourful amphibian, its occurrence in protected areas, such as the Parque Nacional da Serra da Canastra in the state of Minas Gerais and Parque Estadual das Furnas do Bom Jesus in the state of São Paulo, may provide it with some level of protection.

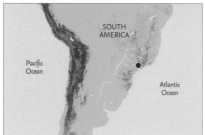

A species is Critically Endangered when it faces an extremely high risk of extinction in the wild, based on measurements of population size and/or geographic range and their trends in the past, present and/or future.

Asian Surf Grass

NOT EVALUATED	DATA DEFICIENT	LEAST CONCERN	NEAR THREATENED	VULNERABLE	❮ ENDANGERED ❯	CRITICALLY ENDANGERED	EXTINCT IN THE WILD	EXTINCT
NE	DD	LC	NT	VU	EN	CR	EW	EX

© Kun-Seop Lee

Asian Surf Grass, Phyllospadix japonicus, is classified as 'Endangered' on the IUCN Red List of Threatened Species™. It lives submerged in the world's oceans, and is found only in the Yellow Sea and the Sea of Japan. Seagrasses are an important coastal habitat that sequester carbon, filter coastal waters, and contribute to ocean productivity.

Although previously widespread, Asian Surf Grass is currently in decline along the coasts of Korea and northern China, with a very limited distribution in Japan. Since the establishment of extensive kelp algal aquaculture, a large proportion of this seagrass species has been permanently removed, sometimes as a result of dynamite fishing in Japan. Shoreline hardening is also a major threat for Asian Surf Grass.

In South Korea, Asian Surf Grass is protected by 'The Marine Ecosystem Conservation and Management Act', and in many areas its distribution falls in Marine Protected Areas. Particularly in Japan and China, this Endangered seagrass species is in need of protection and restoration.

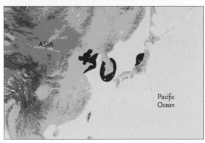

A species is Endangered when it faces a very high risk of extinction in the wild, based on measurements of population size and/or geographic range and their trends in the past, present and/or future.

Rzedowski's Pine

Pinus rzedowskii

NOT EVALUATED	DATA DEFICIENT	LEAST CONCERN	NEAR THREATENED	VULNERABLE	⟨ ENDANGERED ⟩	CRITICALLY ENDANGERED	EXTINCT IN THE WILD	EXTINCT
NE	DD	LC	NT	VU	EN	CR	EW	EX

Rzedowski's Pine, Pinus rzedowskii, is listed as 'Endangered' on the IUCN Red List of Threatened Species™. Only discovered in 1969, this extremely rare and unusual pine is restricted to three distinct populations in western Mexico.

Rzedowski's Pine trees are respected as being unique, and are thought to be safe from exploitation, however fire presents a major threat to this species' survival.

Since this pine has not received formal protection in its native habitat, the three remnant populations of this species need urgent protection and vigilance against forest fires. The establishment of small forestry plantations using seeds from the different populations, together with the species' cultivation in botanical gardens, has been proposed as an additional security measure.

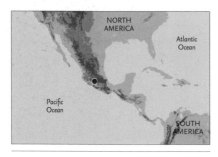

A species is Endangered when it faces a very high risk of extinction in the wild, based on measurements of population size and/or geographic range and their trends in the past, present and/or future.

Golden Vizcacha Rat

Pipanacoctomys aureus

NOT EVALUATED	DATA DEFICIENT	LEAST CONCERN	NEAR THREATENED	VULNERABLE	ENDANGERED	CRITICALLY ENDANGERED	EXTINCT IN THE WILD	EXTINCT
NE	DD	LC	NT	VU	EN	CR	EW	EX

The **Golden Vizcacha Rat**, Pipanacoctomys aureus, is listed as 'Critically Endangered' on the IUCN Red List of Threatened Species™. It is known only from one small area at Salar de Pipanaco in the Catamarca province, Argentina, where it inhabits a narrow band of halophytic (salt-tolerant) plant habitat, and specializes in feeding on the halophytic plants.

The main threat to the Golden Vizcacha Rat is habitat loss due to the expansion of olive plantations. Its global population is confined to a single location, and occupies a total area of less than ten square kilometres. Therefore any threats at this site have the potential to wipe out the entire species.

There are no known specific conservation measures currently in place for the Golden Vizcacha Rat, and the species does not occur in any protected areas. With such a small and declining area of habitat, urgent conservation action is likely to be needed if the species is to be saved from extinction.

© RUBEN BARQUEZ

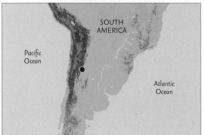

A species is Critically Endangered when it faces an extremely high risk of extinction in the wild, based on measurements of population size and/or geographic range and their trends in the past, present and/or future.

Philippine Eagle

NOT EVALUATED	DATA DEFICIENT	LEAST CONCERN	NEAR THREATENED	VULNERABLE	ENDANGERED	CRITICALLY ENDANGERED	EXTINCT IN THE WILD	EXTINCT
NE	DD	LC	NT	VU	EN	CR	EW	EX

The **Philippine Eagle**, Pithecophaga jefferyi, is listed as 'Critically Endangered' on the IUCN Red List of Threatened Species™. It is found in the Philippines on the larger islands of Luzon, Samar, Leyte and Mindanao. It is the world's second-largest forest eagle and the most endangered of all raptors. Today, there are estimated to be fewer than 500 of these eagles left in the wild.

The Philippine Eagle feeds mainly on flying lemurs, palm civets and monkeys, hence the alternative common name of 'Monkey-eating Eagle'. The number of these majestic birds has seen a steep decline, primarily due to habitat destruction. Since the 1960s, vast tracts of tropical forest have been cleared for commercial development, cultivation and mining activities. Hunting also poses an additional threat, as many local people mistakenly think the eagles take their chickens as prey.

Law in the Philippines protects this eagle, as does CITES (the Convention on International Trade in Endangered Species). A major captive breeding programme is underway in Mindanao, but the key conservation need is to prevent any further forest loss within the range of this species.

Cuvier's Hutia

Plagiodontia aedium

NOT EVALUATED	DATA DEFICIENT	LEAST CONCERN	NEAR THREATENED	VULNERABLE	⟨ ENDANGERED ⟩	CRITICALLY ENDANGERED	EXTINCT IN THE WILD	EXTINCT
NE	DD	LC	NT	VU	EN	CR	EW	EX

© Eladio Fernandez

Cuvier's Hutia, *Plagiodontia aedium*, is listed as 'Endangered' on the IUCN Red List of Threatened Species™. Also known as the Hispaniolan Hutia, this large, mostly arboreal (spending most of its time in trees), herbivorous rodent is one of just two remaining land mammals endemic to the island of Hispaniola (Dominican Republic and Haiti), but very little is known about its populations or biology.

Depending on the area it lives in, Cuvier's Hutia is under threat from a combination of habitat loss, hunting for food, persecution as a crop pest, and predation by introduced dogs, cats, rats and mongooses. The human population on Hispaniola is increasing, putting escalating pressure on the remaining forest, and Cuvier's Hutia has already been lost from much of its former range.

Although Cuvier's Hutia occurs in a number of protected areas, habitat loss is reported to be continuing within these. Further research is needed into this poorly known rodent, and suitable areas of habitat need to be properly protected. A recently initiated project is conducting field research to assess the species' status, threats and specific conservation requirements.

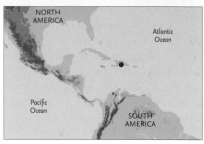

A species is Endangered when it faces a very high risk of extinction in the wild, based on measurements of population size and/or geographic range and their trends in the past, present and/or future.

Pino Plateado

Plantago bismarckii

NOT EVALUATED	DATA DEFICIENT	LEAST CONCERN	NEAR THREATENED	VULNERABLE	**‹ ENDANGERED ›**	CRITICALLY ENDANGERED	EXTINCT IN THE WILD	EXTINCT
NE	DD	LC	NT	VU	EN	CR	EW	EX

The **Pino Plateado** (Silver Pine), *Plantago bismarckii*, is a small gregarious shrub, generally considered to be 'Endangered', although it has not yet been submitted for formal categorization on the IUCN Red List of Threatened Species™. It has long been considered as the flagship plant species of the Sierra de la Ventana mountain range.

Pino Plateado is restricted in range to a small area, and the best stands of this species are protected in the Parque Provincial Ernesto Tornquist. Due to its beautiful silver colour and its shape, it is a distinctive component of the local flora of this area, which is considered to be an island of biodiversity within the Pampas region in east-central Argentina. Its survival is threatened by an ever-increasing use of the area for tourism developments, as well as intense grazing by domestic cattle outside the protected area.

There have been no field studies relating to its population tendency, and little is known about its reproductive biology. Attempts made to cultivate this species outside of its natural area of occurrence have been unsuccessful.

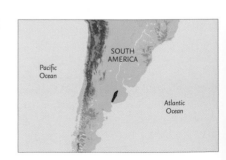

Platanista gangetica

NOT EVALUATED	DATA DEFICIENT	LEAST CONCERN	NEAR THREATENED	VULNERABLE	‹ ENDANGERED ›	CRITICALLY ENDANGERED	EXTINCT IN THE WILD	EXTINCT
NE	DD	LC	NT	VU	EN	CR	EW	EX

© ELISABETH & RUBAIYAT MANSUR/BCDP/WCS

The **South Asian River Dolphin**, *Platanista gangetica*, is listed as 'Endangered' on the IUCN Red List of Threatened Species™. An inhabitant of the Indus, Ganges, Meghna Brahmaputra, and Karnaphuli river systems of the Asian subcontinent, this species has extremely poor eyesight and relies almost entirely on echolocation to navigate and locate prey.

A major threat to the South Asian River Dolphin has been the extensive damming of rivers for irrigation, flood control and electricity generation. Such damming isolates populations, impedes seasonal movements and reduces dry-season flows in the river courses. Other threats include entanglement in fishing gear, chemical pollution, boat traffic, and hunting in a few areas.

This species is legally protected in all the countries it lives in and occurs in a number of national parks and other designated areas, including dolphin reserves or sanctuaries, where at least nominal enforcement takes place. In Pakistan, the enforcement of regulations prohibiting dolphin hunting prevented a rapid population decline in the Indus during the early 1970s.

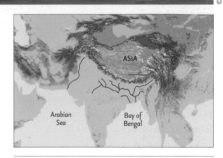

A species is Endangered when it faces a very high risk of extinction in the wild, based on measurements of population size and/or geographic range and their trends in the past, present and/or future.

Platycnemis pembipes

NOT EVALUATED	DATA DEFICIENT	LEAST CONCERN	NEAR THREATENED	VULNERABLE	ENDANGERED	CRITICALLY ENDANGERED	EXTINCT IN THE WILD	EXTINCT
NE	DD	LC	NT	VU	EN	CR	EW	EX

The **Pemba Featherleg**, Platycnemis pembipes, is classified as 'Critically Endangered' on the IUCN Red List of Threatened Species™. This fragile black and white damselfly, named for its white quill-like shins, was first discovered in 2001 on the island of Pemba, just off the coast of Tanzania.

Remarkably, the nearest relative of the Pemba Featherleg occurs on Madagascar, 1,000 km away. Although the species might have reached Pemba from afar aided by strong monsoons, recent studies suggest that it may be the survivor of an ancient African fauna that is now otherwise confined to Madagascar. The Pemba Featherleg only inhabits the single stream flowing through Pemba's last scrap of forest. The island was forest-covered until clearing began for cash crop plantations (cloves and cardamon), and now just a few square kilometres of forest remain.

The Ngezi Forest attained Forest Reserve status in 1959, however the Pemba Featherleg will certainly become extinct if the remaining forest on the island is not conserved.

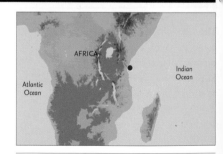

*A species is **Critically Endangered** when it faces an extremely high risk of extinction in the wild, based on measurements of population size and/or geographic range and their trends in the past, present and/or future.*

NOT EVALUATED	DATA DEFICIENT	LEAST CONCERN	NEAR THREATENED	‹ VULNERABLE ›	ENDANGERED	CRITICALLY ENDANGERED	EXTINCT IN THE WILD	EXTINCT
NE	DD	LC	NT	VU	EN	CR	EW	EX

© VIOLA CLAUSNITZER

The **Golden Dancing-jewel**, *Platycypha auripes*, is classified as 'Vulnerable' on the IUCN Red List of Threatened Species™. This species owes its name to its golden colour and to the dancing mating flights that males perform when a female is present. The inflated and coloured legs are used to attract females and to frighten away other males.

The Golden Dancing-jewel is found along clear and fast-running forest streams and rivers in the Eastern Arc Mountains of Tanzania (e.g. Usambara, Ulugur and Udzungwa). The forests are highly fragmented, and suffer from deforestation and water pollution by the growing human population in and around them.

Due to the forest destruction in the Eastern Arc Mountains, often only forests on hill-tops remain which lack suitable breeding habitats for the Golden Dancing-jewel. Currently only parts of the East Usambara and the Udzungwa Mountains experience some kind of protection.

A species is Vulnerable when it faces a high risk of extinction in the wild, based on measurements of population size and/or geographic range and their trends in the past, present and/or future.

Styrian Golden Grasshopper

Podismopsis styriaca

NOT EVALUATED	DATA DEFICIENT	LEAST CONCERN	NEAR THREATENED	VULNERABLE	ENDANGERED	CRITICALLY ENDANGERED	EXTINCT IN THE WILD	EXTINCT
NE	DD	LC	NT	VU	EN	CR	EW	EX

The **Styrian Golden Grasshopper**, Podismopsis styriaca, has not yet been officially classified on the IUCN Red List of Threatened Species™, however it has a provisional assessment of 'Critically Endangered'. It is an endemic species of the Eastern Alps, occuring only in a small area on the eastern slopes of the Zirbitzkogel mountain in Styria, Austria. The species was only recently discovered in 2007 and generally inhabits more moist and sunny heathland habitats.

In central and southeastern Europe, Golden Grasshoppers (Genus Podismopsis) consist of four similar species (P. styriaca, P. relicta, P. transsylvanica, P. keisti), all of which are endemic to small ranges in alpine habitats. The Styrian Golden Grasshopper is characterized by its male genitalia, a short wing length and conspicuous song.

The Zirbitzkogel has an altitude of 2,396 m and is known to have been ice-free during the last glacial period. It supports a rich endemic fauna. The area is protected by the Natura 2000 network of the European Union, but is almost certainly threatened by climate change.

A species is *Critically Endangered* when it faces an extremely high risk of extinction in the wild, based on measurements of population size and/or geographic range and their trends in the past, present and/or future.

NOT EVALUATED	DATA DEFICIENT	LEAST CONCERN	NEAR THREATENED	VULNERABLE	⟨ ENDANGERED ⟩	CRITICALLY ENDANGERED	EXTINCT IN THE WILD	EXTINCT
NE	DD	LC	NT	VU	EN	CR	EW	EX

© ALJOS FARJON

Podocarpus nakaii is listed as 'Endangered' on the IUCN Red List of Threatened Species™. This conifer is endemic to Taiwan, where it is restricted to a small area in the centre of the island, in Nantou County, occurring in broad-leaved evergreen forests at elevations of up to 1,000 metres. It is a medium-sized tree with a relatively straight, slender trunk, blade-like leaves that are concentrated near the tips of the branches, and modified, berry-like seed cones.

Little is known about the threats to this conifer, but it is likely to be impacted by increasing human settlement, agriculture, and forest management activities. The species' small and isolated populations may also put it at increased risk of extinction.

There are no specific conservation measures in place for Podocarpus nakaii. Further research may be needed into the species and the threats it faces before any appropriate actions can be taken to protect it.

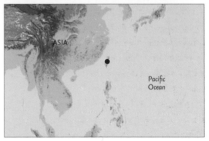

A species is Endangered when it faces a very high risk of extinction in the wild, based on measurements of population size and/or geographic range and their trends in the past, present and/or future.

*Poecilotheria
metallica*

NOT EVALUATED	DATA DEFICIENT	LEAST CONCERN	NEAR THREATENED	VULNERABLE	ENDANGERED	CRITICALLY ENDANGERED	EXTINCT IN THE WILD	EXTINCT
NE	DD	LC	NT	VU	EN	CR	EW	EX

The **Peacock Parachute Spider**, *Poecilotheria metallica*, is listed as 'Critically Endangered' on the IUCN Red List of Threatened Species™. This species is found in a single severely-fragmented area of forest in the Indian State of Andhra Pradesh, between Nandyal and Giddalur. It was in this location that the Peacock Parachute Spider was rediscovered in 2001 after 102 years of not being seen.

This spider is thought to inhabit a range of less than 100 km². This small area is undergoing severe habitat loss and degradation, as a result of lopping for firewood and cutting for timber. An additional threat to the species is collection by international pet traders, which could have an impact on the population. Together, these threats may lead to the extinction of the Peacock Parachute Spider in its single known location in the near future.

It is not known whether or not this species occurs in other parts of the Eastern Ghats. Extensive field surveys are required in order to understand the distribution of this species. Threats other than collection from the wild, such as habitat loss, are yet to be fully understood.

A species is Critically Endangered when it faces an extremely high risk of extinction in the wild, based on measurements of population size and/or geographic range and their trends in the past, present and/or future.

Frigate Island Giant Tenebrionid Beetle

NOT EVALUATED	DATA DEFICIENT	LEAST CONCERN	NEAR THREATENED	VULNERABLE	ENDANGERED	CRITICALLY ENDANGERED	EXTINCT IN THE WILD	EXTINCT
NE	DD	LC	NT	VU	EN	CR	EW	EX

The **Frigate Island Giant Tenebrionid Beetle**, *Polposipus herculeanus*, is classified as 'Critically Endangered' on the IUCN Red List of Threatened Species™. It is one of the largest beetles in the western Indian Ocean and is now only known from Frégate Island, Seychelles, where it is found mainly on the Sandragon Tree, *Pterocarpus indicus*.

The Frigate Island Giant Tenebrionid Beetle survived on Frégate, despite most of the island's native vegetation being replaced by introduced trees. All the adults are flightless, making it almost impossible for them to recolonize their former range. Rats invaded Frégate in 1996, and were swarming over the island by 2000, posing a possible threat to the survival of the beetle.

A captive breeding programme was established at the Zoological Society of London to prevent the species from becoming extinct. A rat eradication programme was also put in place, and Frégate was declared rat-free in 2001. Since then, a new threat has appeared as a result of the deaths of most of the Sandragon Trees due to a fungal infection. However, habitat restoration work has increased the island's forest diversity, enabling the beetle to move onto other native plant species.

© JULIE GANE

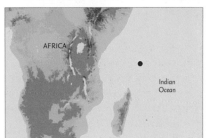

AFRICA

Indian Ocean

A species is Critically Endangered when it faces an extremely high risk of extinction in the wild, based on measurements of population size and/or geographic range and their trends in the past, present and/or future.

NOT EVALUATED	DATA DEFICIENT	LEAST CONCERN	NEAR THREATENED	VULNERABLE	ENDANGERED	CRITICALLY ENDANGERED	EXTINCT IN THE WILD	EXTINCT
NE	DD	LC	NT	VU	EN	CR	EW	EX

The **Tahiti Monarch**, *Pomarea nigra*, is listed as 'Critically Endangered' on the IUCN Red List of Threatened Species™. Restricted to the west coast of Tahiti, this small flycatcher numbers fewer than 50 individuals – just 23 were counted in 2009.

This species was rare throughout the last century, though the reasons for this are not entirely clear. The botanical pest *Miconia calvescens* appears to have played a big part by seriously reducing habitat quality and extent. Moreover, the introduced Black Rat preys on eggs and chicks, and the introduced Red-vented Bulbul and

Common Myna may also be competing with the Tahiti Monarch for resources. With such a small range and tiny population, the Tahiti Monarch is particularly vulnerable to any chance events, such as hurricanes.

Rat control has allowed Tahiti Monarch numbers to start recovering, while further planned conservation measures include improving habitat quality (for example, by removing invasive plants), controlling introduced birds, and initiating a captive breeding programme. Hopefully, these efforts will allow the fragile population to continue its initial recovery.

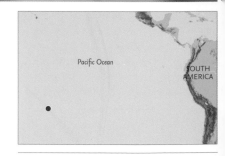

A species is *Critically Endangered* when it faces an extremely high risk of extinction in the wild, based on measurements of population size and/or geographic range and their trends in the past, present and/or future.

Pygmy Hog

Porcula salvania

NOT EVALUATED	DATA DEFICIENT	LEAST CONCERN	NEAR THREATENED	VULNERABLE	ENDANGERED	CRITICALLY ENDANGERED	EXTINCT IN THE WILD	EXTINCT
NE	DD	LC	NT	VU	EN	CR	EW	EX

© Goutam Narayan

Once widely distributed across the alluvial grassland plains in the Himalayan foothills, the **Pygmy Hog**, *Porcula salvania*, declined so severely that by the 1960s it was presumed extinct. Today, one small wild population survives in the tallgrass habitats of Manas National Park, northeast India. It is the world's smallest wild pig, not much taller than a domestic cat, and the most threatened, classified as 'Critically Endangered' on the IUCN Red List of Threatened Species™.

Tall grasslands, being very rich in nutrients, are highly suitable for cultivation and, due to a rapidly growing human population, few now remain. Indiscriminate burning of grass and livestock grazing adds further pressure to the Pygmy Hog's habitat.

Initiated in 1995, the Pygmy Hog Conservation Programme aims to restore the species in historically occupied sites in northeast India through captive breeding and reintroduction, and improved management of the Manas National Park. To date, over 100 individuals have been bred in captivity, and last year 30 Pygmy Hogs were released into the Sonai Rupai Wildlife Sanctuary.

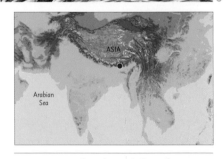

A species is *Critically Endangered* when it faces an extremely high risk of extinction in the wild, based on measurements of population size and/or geographic range and their trends in the past, present and/or future.

NOT EVALUATED	DATA DEFICIENT	LEAST CONCERN	NEAR THREATENED	VULNERABLE	ENDANGERED	CRITICALLY ENDANGERED	EXTINCT IN THE WILD	EXTINCT
NE	DD	LC	NT	VU	EN	CR	EW	EX

RED LIST

© J. TAMELANDER

A species is **Least Concern** when it does not qualify for Critically Endangered, Endangered, Vulnerable or Near Threatened. Widespread and abundant species are included in this category.

Atlantic Ocean

Pacific Ocean

Indian Ocean

Southern Ocean

Porites lutea is listed as 'Least Concern' on the IUCN Red List of Threatened Species™. This coral is found in the Indian and Pacific Oceans, including the Red Sea, Gulf of Aden and The Gulf. Porites species form some of the largest of all coral colonies, with some reaching an incredible eight metres in height.

This species is widespread and common, and is one of the most resilient to stress, such as high temperatures, sedimentation and poor water quality. However, it faces the same threats as many other corals, including disease, human activities and development, invasive species and pollution. Porites corals are also heavily collected for the aquarium trade. With the predicted effects of climate change and ocean acidification, it will be important to regularly reassess the status of this species.

Porites lutea is listed on Appendix II of CITES, and also occurs in several Marine Protected Areas. Management measures are needed to control harvesting for the aquarium trade, while other recommended actions include further research, the expansion of protected areas, and the control of pathogens and parasites.

NOT EVALUATED	DATA DEFICIENT	LEAST CONCERN	NEAR THREATENED	⟨ VULNERABLE ⟩	ENDANGERED	CRITICALLY ENDANGERED	EXTINCT IN THE WILD	EXTINCT
NE	DD	LC	NT	VU	EN	CR	EW	EX

© David Minter

Poronia punctata has not been officially assessed for the IUCN Red List of Threatened Species™, but a provisional assessment of its status is thought to be 'Vulnerable'. It is a microscopic fungus, found exclusively on old dung, usually of donkeys and horses, and occasionally of cattle and elephants. It produces a greyish disc about 15 mm in diameter on the surface of the dung. Many individual fruit-bodies are embedded in the disc, each with an exit channel for its spores. The tops of these channels appear as tiny black points on the disc surface.

Where vehicles have replaced donkeys and horses, populations have declined, making this fungus extremely rare in some parts of the world, particularly Europe. To compete with fungi already present on old dung, Poronia punctata produces antibiotics. Unfortunately, it too is sensitive to non-natural products, such as additives in the food of animals producing the dung, which is an additional factor in its decline.

The fungus is now strongly associated with herbivores on unimproved grasslands, particularly hay meadows, and its conservation therefore requires additive-free food for those herbivores, and protection of the unimproved grasslands where they live.

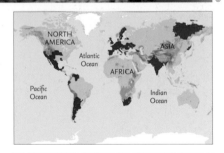

A species is Vulnerable when it faces a high risk of extinction in the wild, based on measurements of population size and/or geographic range and their trends in the past, present and/or future.

Blue River Crab

Potamonautes lividus

The **Blue River Crab**, *Potamonautes lividus*, is listed as 'Vulnerable' on the IUCN Red List of Threatened Species™. This medium-sized crab is endemic to swamp forests in northeastern Kwa-Zulu Natal, South Africa. Individuals inhabit u-shaped burrows which they dig among vegetation in the spongy peat soil. It is terrestrial at night and during rainstorms, when it leaves its burrow to feed on land.

The Blue River Crab is found in isolated patches of wetlands, which are being drained to allow the expansion of human settlement and agriculture. The fragmentation of these areas makes this species particularly vulnerable, as individuals are unlikely to be able to move on to other suitable habitat.

Although populations of this crab are isolated, it is found in three protected areas in northeastern Kwa-Zulu Natal: the Mapelane Nature Reserve, the Mkuze Game Reserve, and the Hluhluwe Game Reserve.

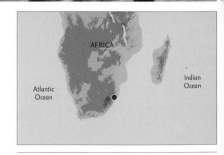

A species is Vulnerable when it faces a high risk of extinction in the wild, based on measurements of population size and/or geographic range and their trends in the past, present and/or future.

Giant Armadillo

Priodontes maximus

© CARLY VYNNE

The **Giant Armadillo**, *Priodontes maximus*, is listed as 'Vulnerable' on the IUCN Red List of Threatened Species™. It is the largest living armadillo species, and occurs east of the Andes in South America, from northern Venezuela and the Guianas south to Paraguay, southern Brazil and northern Argentina.

Although still widespread, the Giant Armadillo is patchily distributed and locally rare, and its population is undergoing a decline. The main threat to the species is hunting for its meat, which is compounded by habitat loss from deforestation. Giant Armadillos are also illegally caught to be sold to animal collectors on the black market, but usually die during transport or whilst in captivity.

The Giant Armadillo occurs in a number of protected areas, and international trade in the species is banned by its listing on Appendix I of CITES. However, illegal hunting continues throughout its range, and measures need to be taken both to decrease hunting pressure and to protect the habitat of this unique mammal.

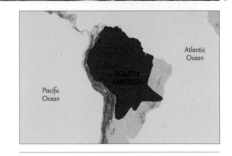

A species is Vulnerable when it faces a high risk of extinction in the wild, based on measurements of population size and/or geographic range and their trends in the past, present and/or future.

Fishing Cat

Prionailurus viverrinus

NOT EVALUATED	DATA DEFICIENT	LEAST CONCERN	NEAR THREATENED	VULNERABLE	⟨ ENDANGERED ⟩	CRITICALLY ENDANGERED	EXTINCT IN THE WILD	EXTINCT
NE	DD	LC	NT	VU	EN	CR	EW	EX

The **Fishing Cat**, *Prionailurus viverrinus*, is listed as 'Endangered' on the IUCN Red List of Threatened Species™. Contradicting the belief that cats dislike water, the Fishing Cat is strongly associated with wetland habitat and frequently enters water to prey upon fish. This cat has been extirpated from much of its former range across South and Southeast Asia, and is now rarely encountered.

The main threat to the Fishing Cat is the destruction and degradation of its wetland habitat due to urban encroachment, drainage for agriculture, pollution and logging. Although not usually perceived as a commercially valuable species, it is vulnerable to snares set for other mammals and its fur may be traded in some Asian markets. This species may also be hunted for food, or in retribution for taking livestock or fish from nets.

With 94 percent of globally significant wetlands in Southeast Asia considered to be under threat, the main conservation priority for the Fishing Cat is to increase the protection afforded to its remaining wetland habitat.

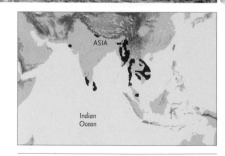

A species is Endangered when it faces a very high risk of extinction in the wild, based on measurements of population size and/or geographic range and their trends in the past, present and/or future.

Crau Plain Grasshopper

NOT EVALUATED	DATA DEFICIENT	LEAST CONCERN	NEAR THREATENED	VULNERABLE	ENDANGERED	CRITICALLY ENDANGERED	EXTINCT IN THE WILD	EXTINCT
NE	DD	LC	NT	VU	EN	CR	EW	EX

The **Crau Plain Grasshopper**, Prionotropis hystrix rhodanica, has not yet been evaluated for the IUCN Red List of Threatened Species™, however it has a provisional assessment of 'Critically Endangered'. It is endemic to the La Crau area – a unique Mediterranean dry steppe in southern France. This species is completely flightless and, therefore, not able to colonize new habitats easily.

The Crau Plain Grasshopper is strongly threatened by the destruction and fragmentation of its habitat resulting from landscape conversion (agriculture and industrial development). Little is known about the population ecology of this species, but the populations seem to be very small and are becoming more and more isolated from each other. In fact, the species has completely disappeared from several sites.

This remarkable grasshopper species is currently the focus of various research projects aimed at establishing conservation management plans. Parts of its natural habitat are protected and managed within the framework of the natural reserve of La Crau, the only zone where this grasshopper can still survive. In order to counteract habitat degradation in other parts of the La Crau, a translocation project is currently being planned.

© Laurent Tatin

A species is Critically Endangered when it faces an extremely high risk of extinction in the wild, based on measurements of population size and/or geographic range and their trends in the past, present and/or future.

Narrowsnout Sawfish

NOT EVALUATED	DATA DEFICIENT	LEAST CONCERN	NEAR THREATENED	VULNERABLE	ENDANGERED	CRITICALLY ENDANGERED	EXTINCT IN THE WILD	EXTINCT
NE	DD	LC	NT	VU	EN	CR	EW	EX

The **Narrowsnout Sawfish**, Pristis zijsron, is listed as 'Critically Endangered' on the IUCN Red List of Threatened Species™. This large and unusual ray has a shark-like body and a long, flattened snout, or saw, which bears numerous pairs of 'teeth'. It is found in the northern Indian Ocean, around Southeast Asia and Australia, and in the western Pacific Ocean.

Like all sawfishes, the Narrowsnout Sawfish is highly vulnerable to both targeted and incidental capture in fisheries, easily becoming entangled in nets due to its large size and the toothed saw. As a result, this species has declined severely in number and range, and may have disappeared completely from some areas.

The Narrowsnout Sawfish may occur in the Great Barrier Reef Marine Park, Australia, and is also a protected species in India. However, strict legal protection is required throughout its range, and further research is needed into the species' biology and populations. Levels of by-catch also need to be monitored if this sawfish is to be better protected.

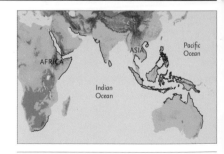

A species is **Critically Endangered** when it faces an extremely high risk of extinction in the wild, based on measurements of population size and/or geographic range and their trends in the past, present and/or future.

NOT EVALUATED	DATA DEFICIENT	LEAST CONCERN	‹ NEAR THREATENED ›	VULNERABLE	ENDANGERED	CRITICALLY ENDANGERED	EXTINCT IN THE WILD	EXTINCT
NE	DD	LC	NT	VU	EN	CR	EW	EX

© Rosser Garrison

Kimmin's Clubtail, Progomphus kimminsi, is classified as 'Near Threatened' on the IUCN Red List of Threatened Species™. This species was described in 1973 but then remained a mystery until its rediscovery in recent years. It is an inhabitant of cloud forest and is restricted to southern Bolivia and northwestern Argentina, where it is found along open streams and rivers and males can be seen patrolling the banks when sunny.

The extent of the occurrence of Kimmin's Clubtail is probably larger than recorded, especially across southern Bolivia which has been scarcely sampled, and it was likely overlooked by non-specialist naturalists due to its pale and cryptic colours. Threats to this species include reduction and fragmentation of habitat by selective logging, and clear-cutting of forest for agriculture and petroleum prospecting (which occurs in at least part of its range).

Searching for further localities is needed in order to conserve Kimmin's Clubtail, as well as studies on its biology and ecology, and monitoring to ensure that its existence does not become threatened by habitat reduction in the future.

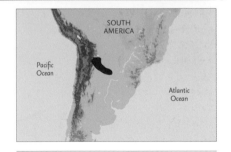

A species is **Near Threatened** when it does not qualify for Critically Endangered, Endangered or Vulnerable now, but is close to qualifying for or is likely to qualify for a threatened category in the near future.

Protium attenuatum

NOT EVALUATED	DATA DEFICIENT	LEAST CONCERN	NEAR THREATENED	VULNERABLE	ENDANGERED	CRITICALLY ENDANGERED	EXTINCT IN THE WILD	EXTINCT
NE	DD	LC	NT	VU	EN	CR	EW	EX

The **Lansan Tree** (Bois L'Encens), Protium attenuatum, is listed as 'Data Deficient' on the IUCN Red List of Threatened Species™. It is a Lesser Antillean endemic, with Saint Lucia appearing to be the main stronghold of this species. This rainforest tree is best known for its aromatic resin, which is widely used as incense in churches and household shrines.

Overexploitation for resin and timber has reportedly decimated this species across most of its restricted range, including Guadeloupe, Dominica, and Saint Vincent and the Grenadines. On Martinique, one of the largest islands within its range, fewer than 50 individuals remain.

More than 60% of Saint Lucians report using Lansan resin, and tapping resin continues to be an important source of income for many households. The Saint Lucia Forestry Department, in partnership with the Global Trees Campaign, is launching a new initiative in 2010 to work with local tappers to improve their harvesting practices and ensure that this important industry is sustainable. The Lansan Tree could become a potent flagship species for the local rainforests, symbolizing the biological, cultural and economic values of a well-managed forest.

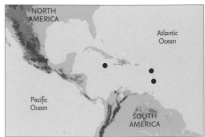

© JENNY DALTRY, FFI

NOT EVALUATED	DATA DEFICIENT	LEAST CONCERN	NEAR THREATENED	VULNERABLE	ENDANGERED	CRITICALLY ENDANGERED	EXTINCT IN THE WILD	EXTINCT
NE	DD	LC	NT	VU	EN	CR	EW	EX

© Rosser W. Garrison

Klugi's Threadtail, Protoneura klugi, has been assessed as 'Data Deficient' on the IUCN Red List of Threatened Species™. This delicate damselfly is known from only a few males found in two localities close to each other in the Amazon forest of northern Peru.

The habitat of Klugi's Threadtail includes rivers in lowland Amazon forest. During a visit to the Research Station of Tiputini, located in the Amazon forest of eastern Ecuador, this species was found inhabiting several narrow streams within the forest. This discovery considerably enlarges the range of distribution of Klugi's Threadtail,

and it is hoped that many other rare species of both dragonfly and damselfly, which are insufficiently known, may be found to be more widely distributed upon further study of this vast and rich forest.

The two known localities of this species are not within protected areas, and are of unknown conservation status. Further information is needed to protect Klugi's Threadtail, including surveys to find more localities and establish distribution range, biological studies, and an evaluation of threats to its continued survival.

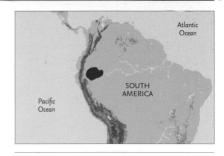

A species is Data Deficient when there is inadequate information to make an assessment of its risk of extinction. Listing of species in this category indicates that more information is required, and acknowledges that future research may show that threatened classification is appropriate.

Darwin's Fox

NOT EVALUATED	DATA DEFICIENT	LEAST CONCERN	NEAR THREATENED	VULNERABLE	ENDANGERED	CRITICALLY ENDANGERED	EXTINCT IN THE WILD	EXTINCT
NE	DD	LC	NT	VU	EN	CR	EW	EX

Darwin's Fox, Pseudalopex fulvipes, is listed as 'Critically Endangered' on the IUCN Red List of Threatened Species™. This small fox has a disjunct distribution, with one population in the coastal mountains on mainland Chile and another 600 km south on Chiloé Island.

With only about 250 individuals remaining, this endemic fox is considered to have the highest extinction risk of any Chilean mammal. Although protected by law, direct persecution by humans continues. While habitat loss due to farming and logging means that the species is pushed into less desirable habitat (pastures and open areas), and closer to human populations, the greatest risk is the presence of domestic dogs in protected areas as potential vectors of disease and fatal attacks on foxes.

There is a local movement to disseminate information about the fox and the threats it faces amongst local schools, dog owners, farmers and loggers. There is also an effort to expand the total protected area on the mainland. Basic research is needed on the species' density, distribution, population genetics, and disease risk/exposure.

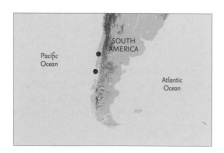

A species is **Critically Endangered** when it faces an **extremely high risk** of extinction in the wild, based on measurements of population size and/or geographic range and their trends in the past, present and/or future.

Western Swamp Turtle

NOT EVALUATED	DATA DEFICIENT	LEAST CONCERN	NEAR THREATENED	VULNERABLE	ENDANGERED	CRITICALLY ENDANGERED	EXTINCT IN THE WILD	EXTINCT
NE	DD	LC	NT	VU	EN	CR	EW	EX

© Gerald Kuchling

The **Western Swamp Turtle**, Pseudemydura umbrina, is listed as 'Critically Endangered' on the IUCN Red List of Threatened Species™. This small freshwater turtle, of an ancient and distinct family, has a prominent spiny neck and its shell grows no larger than 15 cm in length. It inhabits the Perth region of Western Australia, where it persists in only two small seasonal marshes, with less than 100 animals surviving. It spends many months aestivating (the stopping or slowing of activity) during the hot dry summer, emerging for a few months to feed and reproduce during the wet season.

Most of the Western Swamp Turtle's original habitat has been lost to agriculture, housing and mining. This species is also at high risk from global warming and increasing aridity impacting its seasonal wetlands.

Understanding the effects of extreme seasonal changes on the turtles' biology and behaviour proved to be the key to a successful captive breeding programme which, together with intensive protection of the remaining wetlands and reintroduction has averted near-certain extinction.

A species is Critically Endangered when it faces an extremely high risk of extinction in the wild, based on measurements of population size and/or geographic range and their trends in the past, present and/or future.

White-shouldered Ibis

Pseudibis davisoni

NOT EVALUATED	DATA DEFICIENT	LEAST CONCERN	NEAR THREATENED	VULNERABLE	ENDANGERED	CRITICALLY ENDANGERED	EXTINCT IN THE WILD	EXTINCT
NE	DD	LC	NT	VU	EN	CR	EW	EX

The **White-shouldered Ibis**, *Pseudibis davisoni*, is listed as 'Critically Endangered' on the IUCN Red List of Threatened Species™. Having suffered dramatic declines over much of its former range during the course of the 20th century, the White-shouldered Ibis is now confined to just a few sites in Vietnam, extreme southern Laos, Kalimantan (Indonesia) and, principally, northern Cambodia, where some 300 birds survive.

Much of the historical decline of the White-shouldered Ibis was due to habitat loss through logging of lowland forest, drainage of wetlands for agriculture, livestock grazing, grass harvesting, development and hunting. With most populations occurring close to human habitation, disturbance, hunting, logging and infrastructural projects are now putting the most pressure on this species.

As one of the most threatened waterbird species in Southeast Asia, concerted efforts are being made to conserve the White-shouldered Ibis, including various public education campaigns and ecotourism projects. Conservationists are hoping to bring the largest remaining populations under effective protection, a measure which is seen as integral to the recovery of this species.

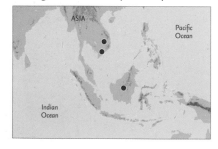

Saola

NOT EVALUATED	DATA DEFICIENT	LEAST CONCERN	NEAR THREATENED	VULNERABLE	ENDANGERED	CRITICALLY ENDANGERED	EXTINCT IN THE WILD	EXTINCT
NE	DD	LC	NT	VU	EN	CR	EW	EX

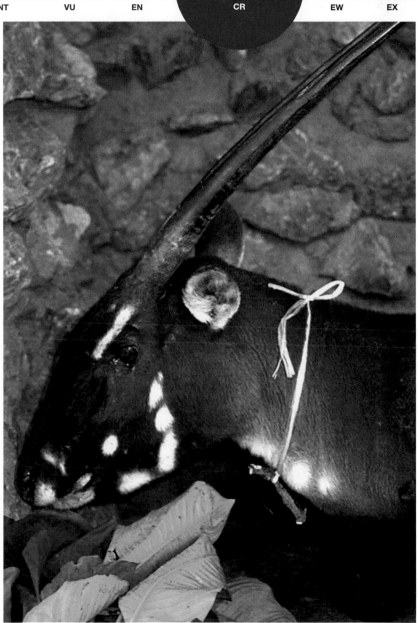

The **Saola**, *Pseudoryx nghetinhensis*, is listed as 'Critically Endangered' on the IUCN Red List of Threatened Species™. Restricted to a narrow area on the border between Vietnam and the Laos, this unusual bovid was discovered as recently as 1992, but is already in grave danger of extinction.

The Saola is under great threat from hunting, with individuals killed for food or snared incidentally in traps set for other animals. This problem is exacerbated by habitat loss, with forests being cleared for agriculture, timber, roads and other development, while logging roads fragment the remaining forest and open it up to hunters. Economic development and human population growth in the region are only likely to increase these threats.

Although the Saola occurs in some protected areas, effective management and conservation of many of these areas has been a problem. The species is legally protected in both Vietnam and Laos, and a number of conservation initiatives are underway. However, the future of this unique but little-known species remains uncertain.

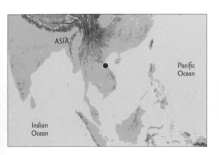

Galvagni's Ground Mantis

Pseudoyersinia andreae

NOT EVALUATED	DATA DEFICIENT	LEAST CONCERN	NEAR THREATENED	⟨ VULNERABLE ⟩	ENDANGERED	CRITICALLY ENDANGERED	EXTINCT IN THE WILD	EXTINCT
NE	DD	LC	NT	VU	EN	CR	EW	EX

© PAOLO FONTANA

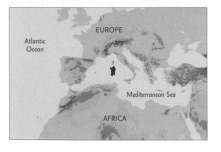

A species is **Vulnerable** when it faces a high risk of extinction in the wild, based on measurements of population size and/or geographic range and their trends in the past, present and/or future.

Galvagni's Ground Mantis, Pseudoyersinia andreae, has not yet been officially classified on the IUCN Red List of Threatened Species™, however it has a provisional listing of 'Vulnerable'. This species was first described in 1976 and is endemic to Sardinia, Italy.

Only a few adult females of this small mantis have been reported, but males have never been found. It is a cryptic insect that occurs in the shrublands of the Sardinian mountains. As is typical of mantids, this species has a small population size. However, in comparison with other genera, its population seems to be extraordinarily small. Only three fragmented localities are known in the central part of Sardinia. The Galvagni's Ground Mantis is threatened as a result of habitat loss through intensive cultivation or conversion to pastures, which has been occurring since the end of 19th century.

There is limited information available on the ecology of this species. Therefore, there is a strong need to study its habitat preferences and life cycle in order to implement conservation measures. Due to its scarcity, Galvagni's Ground Mantis is difficult to protect.

NOT EVALUATED	DATA DEFICIENT	LEAST CONCERN	NEAR THREATENED	VULNERABLE	⟨ ENDANGERED ⟩	CRITICALLY ENDANGERED	EXTINCT IN THE WILD	EXTINCT
NE	DD	LC	NT	VU	EN	CR	EW	EX

© Robert Meyers http://seaclicks.com

The **Banggai Cardinalfish**, *Pterapogon kauderni*, is classified as 'Endangered' on the IUCN Red List of Threatened Species™. This species has a very limited range and is only found in a small island chain known as the Banggai Archipelago, in eastern Indonesia.

The Banggai Cardinalfish has a striking black and white colouration and ornate fins, making it a desirable species for aquarium hobbyists. As this popularity increases, so has the demand; the fish are collected all over the archipelago and are then shipped around the world to various aquarium dealers. It is estimated that currently over 9,000 individuals are being taken off reefs and put into the aquarium trade each year.

Currently there are captive breeding programmes to help supply the aquarium industry with non-wild-caught specimens. There are also several organizations that have regulated the trade of this species and who advise responsible aquarists not to purchase wild-caught individuals. Although the aquarium trade has mostly been detrimental to this species, it has also inadvertently expanded its range by introducing the fish to new habitats such as Lembeh, Indonesia.

A species is Endangered when it faces a very high risk of extinction in the wild, based on measurements of population size and/or geographic range and their trends in the past, present and/or future.

Livingstone's Flying Fox

Pteropus livingstonii

NOT EVALUATED	DATA DEFICIENT	LEAST CONCERN	NEAR THREATENED	VULNERABLE	‹ ENDANGERED ›	CRITICALLY ENDANGERED	EXTINCT IN THE WILD	EXTINCT
NE	DD	LC	NT	VU	EN	CR	EW	EX

Livingstone's Flying Fox, Pteropus livingstonii, is one of the largest and most threatened bats in the world and is classified as 'Endangered' on the IUCN Red List of Threatened Species™. Around 1,000 individuals remain in the forests clinging onto the precipitous mountain slopes of the western Indian Ocean islands of Anjouan and Mohéli in the Union of the Comoros. Flying Foxes have a vital role in the dynamics of forest ecosystems as they pollinate flowers and disperse seeds.

The Comoros have lost most of their forests and still suffer among the world's highest deforestation rates. The surviving Livingstone's Flying Foxes are highly susceptible to further forest loss, destruction of their habitual roosts and the impacts of natural disasters such as cyclones.

A project recently started to establish the first forest protected areas on these islands to conserve Livingstone's Flying Fox and other wildlife. A captive breeding programme in a number of UK zoos provides a back-up population of this species should the worst happen in the wild.

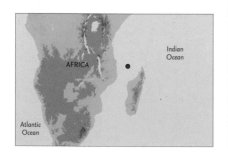

© JAMES MORGAN

NOT EVALUATED	DATA DEFICIENT	LEAST CONCERN	NEAR THREATENED	⟨ VULNERABLE ⟩	ENDANGERED	CRITICALLY ENDANGERED	EXTINCT IN THE WILD	EXTINCT
NE	DD	LC	NT	VU	EN	CR	EW	EX

© JÖRG FREYHOF

Denison's Barb, *Puntius denisonii*, is listed as 'Vulnerable' on the IUCN Red List of Threatened Species™. This attractive freshwater fish is endemic to the Western Ghats, India, and is one of the most popular ornamental fishes in international trade.

Owing to its popularity, Denison's Barb has been exploited at unsustainable levels since it was first exported from India in 1996. In the absence of catch limits, the population has decreased at an alarming rate of around 70 percent in recent years, and consequently the species now has a highly fragmented distribution, and may occur in as few as four rivers. The species' ability to recover from this exploitation may be limited by low genetic diversity and a sex ratio highly skewed towards males.

The main priority for this species is to limit catch efforts in local fisheries and develop sustainable harvesting methods. Various management plans also need to be explored, including no-take zones and protected areas, while captive-breeding stocks are also being established, which could potentially reduce the impact of trade on wild populations.

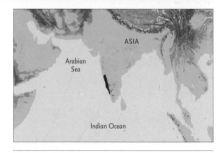

A species is Vulnerable when it faces a high risk of extinction in the wild, based on measurements of population size and/or geographic range and their trends in the past, present and/or future.

Azores Bullfinch

Pyrrhula murina

NOT EVALUATED	DATA DEFICIENT	LEAST CONCERN	NEAR THREATENED	VULNERABLE	‹ ENDANGERED ›	CRITICALLY ENDANGERED	EXTINCT IN THE WILD	EXTINCT
NE	DD	LC	NT	VU	EN	CR	EW	EX

The **Azores Bullfinch**, *Pyrrhula murina*, is currently listed as 'Endangered' on the IUCN Red List of Threatened Species™, having recently been down-listed from 'Critically Endangered' due to successful conservation efforts which focused on this species and its habitat. This rare finch is confined to the east of the island of São Miguel, in the Azores, Portugal.

The decline of the Azores Bullfinch is believed to have resulted from widespread forest clearance for agriculture and forestry plantations, while the spread of invasive exotic plants has further degraded the natural vegetation, reducing the species' food supply. Predation by introduced cats and rats may also be a problem. This species has a restricted range, making it vulnerable to any chance events, and its reduced numbers may also lead to problems with inbreeding.

The Azores Bullfinch and its habitat are legally protected, population monitoring is ongoing, and efforts are being made to control exotic plants and to restore and enlarge the remaining areas of native vegetation. However, this is a true conservation success story, with an increase from 250 individuals only a few years ago to now over 1,000.

A species is **Endangered** when it faces a *very high risk* of extinction in the wild, based on measurements of population size and/or geographic range and their trends in the past, present and/or future.

Hinton's Oak

Quercus hintonii

NOT EVALUATED	DATA DEFICIENT	LEAST CONCERN	NEAR THREATENED	VULNERABLE	ENDANGERED	CRITICALLY ENDANGERED	EXTINCT IN THE WILD	EXTINCT
NE	DD	LC	NT	VU	EN	CR	EW	EX

© JUAN PABLO MOREIRAS, FFI

Hinton's Oak, Quercus hintonii, is listed as 'Critically Endangered' on the IUCN's Red List of Threatened Species™. It is found in sub-montane to montane dry forest in Mexico.

Hinton's Oak has a restricted habitat and is thought to have strict altitudinal requirements. It has become threatened in recent years due to the serious destruction and reduction in size of its habitat. The wood has a variety of local uses, being part of the traditional culture of Tejupilco people, for tool handles, beams and fencing poles, but primarily for firewood. The wood is also traditionally used to bake 'las finas' bread, the characteristic taste of which is imparted by the smoke.

Conservation measures include the involvement of local authorities and landowners, training on plant propagation, field research, and the development of an education campaign. The Global Trees Campaign conducted surveys on the species, with collaborative research between Mexican experts from the University of Puebla and staff from the Sir Harold Hillier Gardens & Arboretum in the UK. This has led to the development of a conservation strategy for Hinton's oak.

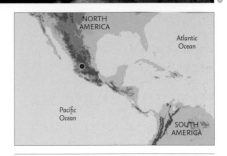

A species is Critically Endangered when it faces an extremely high risk of extinction in the wild, based on measurements of population size and/or geographic range and their trends in the past, present and/or future.

Yangtze Giant Softshell

Rafetus swinhoei

NOT EVALUATED	DATA DEFICIENT	LEAST CONCERN	NEAR THREATENED	VULNERABLE	ENDANGERED	CRITICALLY ENDANGERED	EXTINCT IN THE WILD	EXTINCT
NE	DD	LC	NT	VU	EN	CR	EW	EX

The **Yangtze Giant Softshell**, *Rafetus swinhoei*, is listed as 'Critically Endangered' on the IUCN Red List of Threatened Species™. This enormous softshell turtle, of which only four surviving individuals are known, can reach 120 kg in size and historically inhabited the Red River of Yunnan and Vietnam, and the lower Yangtze floodplain.

Although worshipped in some areas, capture for consumption, wetland destruction, and water pollution, have all severely impacted upon this species. Two wild males remain in different lakes in Vietnam, and a male and female survive in the Suzhou and Changsha zoos in China; these were brought together and have produced several egg clutches over the last two years. Though none have hatched yet, captive breeding efforts continue. The two wild animals are closely monitored; one was saved after the dam impounding its lake broke during flooding and it was captured downriver and returned to its repaired lake.

Field surveys to locate additional animals in the wild continue. The species' prospects for survival remain dire, but international conservation efforts are beginning to turn the tide.

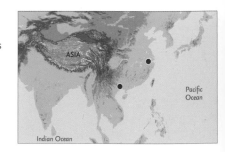

Café Marron

NOT EVALUATED	DATA DEFICIENT	LEAST CONCERN	NEAR THREATENED	VULNERABLE	ENDANGERED	CRITICALLY ENDANGERED	EXTINCT IN THE WILD	EXTINCT
NE	DD	LC	NT	VU	EN	CR	EW	EX

© Vincent Florens

Café Marron – the local name for 'Wild Coffee' – Ramosmania rodriguesii, is a shrubby relative of the coffee plant found only on the Indian Ocean island of Rodrigues. It is classified as 'Critically Endangered' on the IUCN Red List of Threatened Species™, being known only from a single plant in the wild which was discovered when the species was thought to be extinct.

Rodrigues Island has suffered rapid and massive deforestation due to its small size and terrain suited to agricultural use. Less than 1% of the original native habitats currently exist, making habitat destruction the overwhelming main cause of decline of Café Marron. While successful propagation saved the species from imminent extinction, the remaining wild plant still requires protection from those who believe it to have medicinal properties.

Café Marron has been successfully propagated at the Royal Botanic Gardens, Kew, and several plants returned to Rodrigues Island where some were reintroduced in the Grande Montagne Nature Reserve as the first step towards reinstating a viable wild population. It is cultivated in several other places.

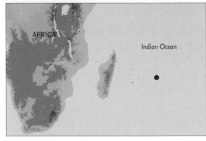

A species is Critically Endangered when it faces an extremely high risk of extinction in the wild, based on measurements of population size and/or geographic range and their trends in the past, present and/or future.

Reindeer

Rangifer tarandus

The **Reindeer**, *Rangifer tarandus*, is listed as 'Least Concern' on the IUCN Red List of Threatened Species™. This species has a circumpolar distribution in the tundra and taiga zones of northern Europe, Siberia and North America. In North America, wild populations remain only in Canada, Alaska, Washington and northern Idaho. In Europe, wild populations have a fragmented distribution in Norway, Finland and Russia.

Herds in northeastern Alaska, north-western Canada and Central Arctic are potentially threatened by onshore petroleum exploration and development, which may displace calving and lead to increased calf mortality. Poaching is a major threat in the Russian Federation.

Reindeer may also be threatened by loss of habitat in Finland and increased disturbance in some areas due to winter sports. The Reindeer is protected under the Bern Convention on the Conservation of European Wildlife and Natural Habitats, which has been signed by most European states. Hunting of Reindeer is strictly controlled in eastern Russia and Norway; however, despite this, poaching continues in Russia. In Finland, semi-domesticated Reindeer and forest Reindeer are separated, to prevent hybridization.

A species is **Least Concern** when it does not qualify for Critically Endangered, Endangered, Vulnerable or Near Threatened. Widespread and abundant species are included in this category.

© SHUTTERSTOCK · ANDREAS GRADIN

Whale Shark  *Rhincodon typus*

NOT EVALUATED	DATA DEFICIENT	LEAST CONCERN	NEAR THREATENED	< VULNERABLE >	ENDANGERED	CRITICALLY ENDANGERED	EXTINCT IN THE WILD	EXTINCT
NE	DD	LC	NT	VU	EN	CR	EW	EX

© Pedro Vieyra, Marine Conservation Society Seychelles

The **Whale Shark**, Rhincodon typus, is listed as 'Vulnerable' on the IUCN Red List of Threatened Species™. Found worldwide in tropical and warm temperate waters, it is the largest fish in the world, yet feeds almost entirely on tiny plankton, crustaceans and small fish.

The Whale Shark has been targeted in many areas for its flesh, liver oil and fins. Although relatively little is known about the biology of this species, its long lifespan and slow reproductive rate, together with a naturally low abundance and highly migratory nature, are likely to increase its vulnerability to overexploitation by fisheries.

Whale Sharks are legally protected in a number of countries, and international trade in the species is regulated by CITES. This huge, charismatic shark now plays an important role in ecotourism in places such as Thailand, South Africa, the Maldives and the Seychelles, providing a valuable source of income for local economies. Although the impacts on the sharks' behaviour are currently unknown, tourism may help to protect the species by giving it more value alive than fished.

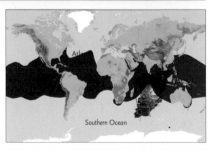

A species is Vulnerable when it faces a high risk of extinction in the wild, based on measurements of population size and/or geographic range and their trends in the past, present and/or future.

Javan Rhinoceros

NOT EVALUATED	DATA DEFICIENT	LEAST CONCERN	NEAR THREATENED	VULNERABLE	ENDANGERED	CRITICALLY ENDANGERED	EXTINCT IN THE WILD	EXTINCT
NE	DD	LC	NT	VU	EN	CR	EW	EX

The **Javan Rhinoceros**, Rhinoceros sondaicus, is listed as 'Critically Endangered' on the IUCN Red List of Threatened Species™. One of the rarest large mammals in the world, it was once widespread across Southeast Asia, but is now ony found in two small areas of Java (Indonesia) and Vietnam. The Vietnamese population numbers only around six individuals, and may not be breeding.

This species' devastating decline has been attributed to overhunting for its horns and other body parts, which are used in traditional Chinese medicine. Habitat loss has also had an impact, and the two tiny remaining populations are threatened by disease and poaching.

The Javan Rhinoceros occurs in two protected areas and is legally protected throughout its range, while international trade in the species is banned under its CITES listing. Recommendations for the species' conservation include captive breeding, as well as moving some of the Javan population into other areas, to reduce the chances of it being wiped out by a single event. However, with so few individuals remaining, the future of this highly endangered mammal is uncertain.

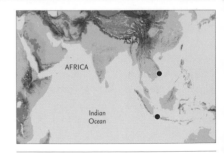

A species is **Critically Endangered** when it faces an extremely high risk of extinction in the wild, based on measurements of population size and/or geographic range and their trends in the past, present and/or future.

Darwin's Frog

Rhinoderma darwinii

NOT EVALUATED	DATA DEFICIENT	LEAST CONCERN	NEAR THREATENED	⟨ VULNERABLE ⟩	ENDANGERED	CRITICALLY ENDANGERED	EXTINCT IN THE WILD	EXTINCT
NE	DD	LC	NT	VU	EN	CR	EW	EX

© JAIME BOSCH

Darwin's Frog, Rhinoderma darwinii, is listed as 'Vulnerable' on the IUCN Red List of Threatened Species™. First discovered by Charles Darwin, Darwin's Frog occurs in the forests of central and southern Chile and Argentina, and has an intriguing system of reproduction in which the male broods the eggs and tadpoles within its vocal sacs.

Darwin's Frog has undergone a worrying decline in recent years, with some populations disappearing entirely. In some parts of its range, these losses can be attributed to habitat loss through deforestation and the replacement of native trees with introduced pine and eucalyptus, as well as to drought. However, in other areas the exact cause of the decline is unknown, but may be linked to global climate change or disease.

Although Darwin's Frog is found in several protected areas, there is a need for improved management of these sites and the protection of more areas. Close population monitoring will also be required in order to help identify the causes of decline in apparently suitable habitat.

A species is **Vulnerable** when it faces a high risk of extinction in the wild, based on measurements of population size and/or geographic range and their trends in the past, present and/or future.

Tonkin Snub-nosed Monkey

Rhinopithecus avunculus

NOT EVALUATED	DATA DEFICIENT	LEAST CONCERN	NEAR THREATENED	VULNERABLE	ENDANGERED	CRITICALLY ENDANGERED	EXTINCT IN THE WILD	EXTINCT
NE	DD	LC	NT	VU	EN	CR	EW	EX

The **Tonkin Snub-nosed Monkey**, Rhinopithecus avunculus, is listed as 'Critically Endangered' on the IUCN Red List of Threatened Species™. The range of this unusual and distinctive primate has been drastically reduced, and it is now known from just five isolated locations within Vietnam.

Due to massive deforestation and intensive hunting, just 200 to 250 individuals of this species remain, and its population is highly fragmented. Although not the main target of hunters, it is still shot when encountered, and is either eaten or used in traditional medicine. Widespread forest destruction has reduced and fragmented its habitat, while the recent development of a hydroelectric project has not only caused further habitat loss, but also an influx of construction workers, resulting in an increased demand for meat and resources.

Despite legal protection, hunting and habitat loss unfortunately continue throughout the Tonkin Snub-nosed Monkey's range. Various conservation efforts are underway, including efforts to raise local awareness, establish patrol groups, undertake further surveys and impose gun controls, but the future of this highly endangered primate still hangs in the balance.

© Cédric Libert

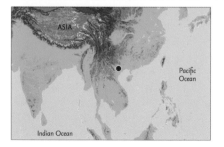

A species is **Critically Endangered** when it faces an extremely high risk of extinction in the wild, based on measurements of population size and/or geographic range and their trends in the past, present and/or future.

NOT EVALUATED	DATA DEFICIENT	LEAST CONCERN	NEAR THREATENED	⟨ VULNERABLE ⟩	ENDANGERED	CRITICALLY ENDANGERED	EXTINCT IN THE WILD	EXTINCT
NE	DD	LC	NT	VU	EN	CR	EW	EX

© Ron DeCloux

The **Javanese Cownose Ray**, *Rhinoptera javanica*, is listed as 'Vulnerable' on the IUCN Red List of Threatened Species™. This species occurs widely throughout the Indo-West Pacific, from South Africa to Japan.

Little is known about the true impact of fisheries on the Javanese Cownose Ray. However, it is likely to be under threat due to its small litter size, its tendency to form large schools, its inshore habitat and hence susceptibility to a wide variety of inshore fishing gear, and the generally intense and unregulated nature of inshore fisheries across most of the species' range.

In order to improve the conservation status of the Javanese Cownose Ray, attempts need to be made to monitor and regulate fisheries throughout its range. In addition, further research into this species' life history, including its growth, longevity, movement patterns, habitat use, reproduction, potential nursery areas, and diet, is necessary in order to develop effective conservation actions.

A species is Vulnerable when it faces a high risk of extinction in the wild, based on measurements of population size and/or geographic range and their trends in the past, present and/or future.

Jerdon's Courser

NOT EVALUATED	DATA DEFICIENT	LEAST CONCERN	NEAR THREATENED	VULNERABLE	ENDANGERED	CRITICALLY ENDANGERED	EXTINCT IN THE WILD	EXTINCT
NE	DD	LC	NT	VU	EN	CR	EW	EX

Jerdon's Courser, Rhinoptilus bitorquatus, is listed as 'Critically Endangered' on the IUCN Red List of Threatened Species™. The single, small and declining population of this poorly known long-legged wading bird can only be found in south-eastern India in rolling scrub forest. It is a largely nocturnal species, and is very rarely seen.

There are thought to be between 50 and 250 Jerdon's Coursers left in the wild. Although relatively little is known about this species, its habitat is under increasing pressure from disturbance and fragmentation. Recent relocation of human populations may pose a serious threat to the remaining habitat and to the birds themselves.

Further research is needed to understand the behaviour, range and potential threats to Jerdon's Courser, but it is so elusive that researchers have difficulty finding birds to study. Its habitat is in a protected area, and lobbying against quarrying and mining activities that threaten existing habitat has proved successful.

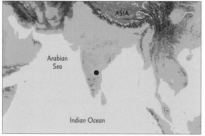

A species is **Critically Endangered** when it faces an extremely high risk of extinction in the wild, based on measurements of population size and/or geographic range and their trends in the past, present and/or future.

Golden-rumped Sengi

Rhynchocyon chrysopygus

RED LIST

NOT EVALUATED	DATA DEFICIENT	LEAST CONCERN	NEAR THREATENED	VULNERABLE	❮ ENDANGERED ❯	CRITICALLY ENDANGERED	EXTINCT IN THE WILD	EXTINCT
NE	DD	LC	NT	VU	EN	CR	EW	EX

© Galen Rathbun, California Academy of Sciences

The **Golden-rumped Sengi**, *Rhynchocyon chrysopygus*, is listed as 'Endangered' on the IUCN Red List of Threatened Species™. Also known as the Golden-rumped Elephantshrew, this species has a large, flexible snout, with which it forages through leaf litter for invertebrates. Its range is along the coast of Kenya, from the Mombasa area north to the mouth of the Tana River. There is no longer any evidence that it crosses the river – instead, a potentially different species is found between the Tana and Somalia.

The Golden-rumped Sengi is severely threatened by habitat destruction along the Kenyan coast, with forests being relentlessly cleared for farming, development and timber collection. Illegal trapping of this species for food also occurs, although current levels are thought to be sustainable.

Although there are no specific conservation measures in place for this species other than a monitoring programme, it does occur within the Arabuko-Sokoke Forest. This partially protected area is the focus of a project to promote long-term conservation through sustainable management and community participation in forest conservation.

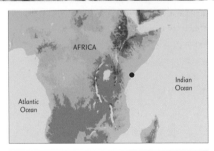

A species is Endangered when it faces a very high risk of extinction in the wild, based on measurements of population size and/or geographic range and their trends in the past, present and/or future.

Grey-faced Sengi

Rhynchocyon udzungwensis

NOT EVALUATED	DATA DEFICIENT	LEAST CONCERN	NEAR THREATENED	⟨ VULNERABLE ⟩	ENDANGERED	CRITICALLY ENDANGERED	EXTINCT IN THE WILD	EXTINCT
NE	DD	LC	NT	VU	EN	CR	EW	EX

The **Grey-faced Sengi**, *Rhynchocyon udzungwensis*, is listed as 'Vulnerable' on the IUCN Red List of Threatened Species™. This unusual mammal is endemic to the Udzungwa Mountains of Tanzania, where it was discovered in 2005 and described in 2008.

Currently, the most significant threat to the Grey-faced Sengi are the occurence of natural and human-caused fires that burn on the surrounding edges of the forests, gradually reducing the extent of their habitat. The growth and expansion of the local human population is only likely to increase the pressures on the forest, while global climate change may potentially further reduce forest cover. This species' restriction to just two locations makes it particularly vulnerable to these threats.

This sengi occurs entirely within the Udzungwa Mountains National Park and Kilombero Nature Reserve. No utilization of animals for food or trade is permitted in these areas, and various conservation activities are underway within the National Park. The Grey-faced Sengi is just one of many endemic species discovered in these mountains in the last decade, highlighting the importance of conserving these ancient forests.

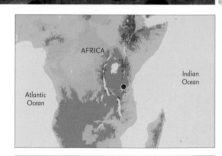

A species is **Vulnerable** when it faces a high risk of extinction in the wild, based on measurements of population size and/or geographic range and their trends in the past, present and/or future.

Madagascan Rousette

Rousettus madagascariensis

NOT EVALUATED	DATA DEFICIENT	LEAST CONCERN	NEAR THREATENED	VULNERABLE	ENDANGERED	CRITICALLY ENDANGERED	EXTINCT IN THE WILD	EXTINCT
NE	DD	LC	NT	VU	EN	CR	EW	EX

© Merlin D. Tuttle, Bat Conservation International, www.batcon.org

The **Madagascan Rousette**, *Rousettus madagascariensis*, is listed as 'Near Threatened' on the IUCN Red List of Threatened Species™. It is the smallest of Madagascar's three endemic fruit bats, and is widespread across much of the island.

The Madagascan Rousette is commonly hunted at its roosts, and is also sometimes killed as a pest of fruit crops. This bat is regarded as a game species under Malagasy law, and only receives nominal protection where it occurs in nature reserves or roosts at sacred sites. A slow reproductive rate makes the species particularly susceptible to overhunting, while deforestation is likely to present a further threat.

Although the Madagascan Rousette occurs in several protected areas, its roost sites are in need of more effective protection both within and outside of reserves. The local NGO Madagasikara Voakajy is actively conserving the species by engaging communities to increase the level of protection at important cave roosts.

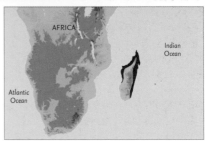

A species is *Near Threatened* when it does not qualify for Critically Endangered, Endangered or Vulnerable now, but is close to qualifying for or is likely to qualify for a threatened category in the near future.

NOT EVALUATED	DATA DEFICIENT	LEAST CONCERN	NEAR THREATENED	VULNERABLE	ENDANGERED	CRITICALLY ENDANGERED	EXTINCT IN THE WILD	EXTINCT
NE	DD	LC	NT	VU	EN	CR	EW	EX

The **Kipunji**, *Rungwecebus kipunji*, is listed as 'Critically Endangered' on the IUCN Red List of Threatened Species™. Endemic to Tanzania, the Kipunji is Africa's most recently discovered primate, having been first found in 2003.

This newly discovered primate is already under serious threat of extinction, numbering only around 1,100 individuals and restricted to just two locations. The Kipunji faces a number of significant threats, including habitat loss through logging and charcoal-making, habitat fragmentation, and illegal hunting for food and in retaliation for crop-raiding. Many subpopulations are small and isolated, and are unlikely to be viable in the long-term.

The Kipunji is an important flagship species for conservation in and around Rungwe-Kitulo, and occurs entirely within protected areas, although only Kitulo National Park has any current management activities, and these remain limited. The effective protection and restoration of its habitat, and in particular of connecting forest 'corridors', together with ongoing population monitoring and public education, will be vital if this unique monkey is to survive.

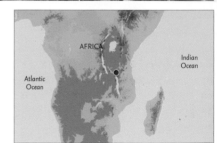

A species is Critically Endangered when it faces an extremely high risk of extinction in the wild, based on measurements of population size and/or geographic range and their trends in the past, present and/or future.

Pied Tamarin

Saguinus bicolor

© ADAIR BRCUGHTON

The **Pied Tamarin** or Brazilian Bare-faced Tamarin, *Saguinus bicolor*, is a small monkey endemic to the Brazilian Amazon, and is classified as 'Endangered' on the IUCN Red List of Threatened Species™. It occurs largely within and around the city of Manaus, in the heart of the Amazon basin, and has one of the smallest ranges of any primate.

The expansion of Manaus has reduced much of the species' habitat to mere fragments which are disappearing rapidly, destroyed by people in search of land and by land-use planning that fails to take environmental needs into account. Tamarins migrating from one tiny patch of forest to another are often electrocuted by power cables or are run over whilst crossing roads.

Translocation of these primates to safer patches of forest is now being implemented to help conserve this species. Seven potential conservation areas for Pied Tamarins have been identified. These areas require protection, as well as the creation of forest corridors to connect them, in order to secure the future of this species in the wild.

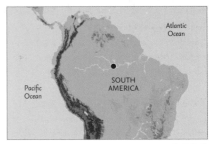

A species is **Endangered** when it faces a very high risk of extinction in the wild, based on measurements of population size and/or geographic range and their trends in the past, present and/or future.

Saiga Antelope

Saiga tatarica

NOT EVALUATED	DATA DEFICIENT	LEAST CONCERN	NEAR THREATENED	VULNERABLE	ENDANGERED	CRITICALLY ENDANGERED	EXTINCT IN THE WILD	EXTINCT
NE	DD	LC	NT	VU	EN	CR	EW	EX

The **Saiga Antelope**, *Saiga tatarica*, is listed as 'Critically Endangered' on the IUCN Red List of Threatened Species™. Saigas have a distinctive appearance, with an unusual swollen nose that filters out airborne dust during the dry summer, and warms cold air before it enters the lungs during the winter. This species is found in isolated populations in southeastern Europe, Central Asia and western Mongolia.

Despite once numbering over one million in the mid-1970s, intense hunting pressure has reduced the global population to some 50,000 individuals. Saiga may be hunted for meat or for their horns, which are highly valued in traditional Chinese medicine due to alleged curative properties. As only males bear horns, selective hunting has distorted the sex ratio, resulting in reproductive failures.

Coordinated conservation measures have now stabilized the decline in Saiga Antelope but hunting has continued, despite being illegal in all countries throughout its range. To conserve this species, the strengthening of anti-poaching laws, proper enforcement of legislation and the extension of protected areas are all crucial.

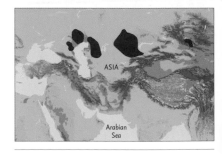

A species is *Critically Endangered* when it faces an extremely high risk of extinction in the wild, based on measurements of population size and/or geographic range and their trends in the past, present and/or future.

Atlantic Salmon

Salmo salar

RED LIST

NOT EVALUATED	DATA DEFICIENT	LEAST CONCERN		NEAR THREATENED	VULNERABLE	ENDANGERED	CRITICALLY ENDANGERED	EXTINCT IN THE WILD	EXTINCT
NE	DD	LC	>	NT	VU	EN	CR	EW	EX

© Michel Roggo

The **Atlantic Salmon**, *Salmo salar*, is listed as 'Least Concern' on the IUCN Red List of Threatened Species™. This species is found throughout the North Atlantic region, where after long marine migrations they home precisely to their natal river to spawn.

The abundance of Atlantic Salmon has declined markedly since the 1970s despite major reductions in fishing effort and other measures. Many river populations are threatened with extinction, particularly throughout their southern range in Europe and North America. Increased natural mortality at sea appears to be a major factor in this decline, potentially associated with climate change. Other threats include river pollution, overfishing and dams. Commercial salmon farms also pose ecological threats, including elevated levels of pathogens, while escaped farm salmon may interbreed with wild salmon.

Significant conservation efforts are underway to help protect the Atlantic Salmon, including those of the inter-governmental North Atlantic Salmon Conservation Organisation. Despite these efforts, abundance remains low and conservation of this iconic fish continues to be a challenge.

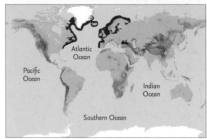

A species is *Least Concern* when it does not qualify for Critically Endangered, Endangered, Vulnerable or Near Threatened. Widespread and abundant species are included in this category.

Red-headed Vulture

Sarcogyps calvus

NOT EVALUATED	DATA DEFICIENT	LEAST CONCERN	NEAR THREATENED	VULNERABLE	ENDANGERED	CRITICALLY ENDANGERED	EXTINCT IN THE WILD	EXTINCT
NE	DD	LC	NT	VU	EN	CR	EW	EX

The **Red-headed Vulture**, *Sarcogyps calvus*, is listed as 'Critically Endangered' on the IUCN Red List of Threatened Species™. Formerly widespread throughout the Indian subcontinent and Southeast Asia, this species has undergone significant declines in both range and population size. It has become uncommon in Nepal, and is rare in Pakistan, the northeast of Bangladesh, Bhutan, Myanmar, Laos, Cambodia and Vietnam. There are currently fewer than 10,000 of these vultures in the wild.

The massive decline in Red-headed Vultures is presumed to be caused by the consumption of livestock treated with the veterinary drug Diclofenac (used in the treatment of inflammation, pain and fever). This drug has been proven to be responsible for the huge declines in populations of vultures of the genus Gyps, causing mortality from kidney failure resulting from visceral gout.

The manufacture of Diclofenac has been banned in India, Nepal and Pakistan; however, it may be years before the use of this drug ends completely in these countries. Red-headed Vultures are being carefully monitored in the Indian subcontinent and Cambodia to ensure that any further declines are detected.

© GANESH H. SHANKAR

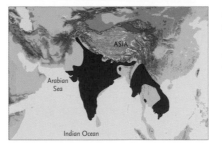

A species is **Critically Endangered** when it faces an extremely high risk of extinction in the wild, based on measurements of population size and/or geographic range and their trends in the past, present and/or future.

Tasmanian Devil

Sarcophilus harrisii

NOT EVALUATED	DATA DEFICIENT	LEAST CONCERN	NEAR THREATENED	VULNERABLE	ENDANGERED	CRITICALLY ENDANGERED	EXTINCT IN THE WILD	EXTINCT
NE	DD	LC	NT	VU	EN	CR	EW	EX

© SHUTTERSTOCK - PATSY A. JACKS

The **Tasmanian Devil**, *Sarcophilus harrisii*, is listed as 'Endangered' on the IUCN Red List of Threatened Species™. Although now found only in Tasmania, this iconic animal formerly occupied much of the Australian mainland, but disappeared at least 400 years ago, and possibly as long as several thousand years ago.

Historically, early European settlers considered the Tasmanian Devil a nuisance that killed poultry. As a consequence, it was intensively persecuted for many years through trapping and poisoning until a protective law was passed in 1941 that saw numbers gradually rise again. Today, the greatest threat to this species is the fatal cancer known as Devil Facial Tumour Disease (DFTD). Since its discovery, Tasmanian Devil numbers have declined by 60%, and localized declines exceed 90% in places where the disease has been present for the longest time.

Considerable effort is being made at the local, national and international level to reverse the fate of the Tasmanian Devil. At the forefront of this is the Save the Tasmanian Devil Program, which aims to monitor the impact of DFTD, develop methods of managing its impact and maintain a disease-free insurance population.

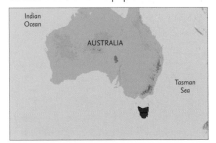

NOT EVALUATED	DATA DEFICIENT	LEAST CONCERN	NEAR THREATENED	< VULNERABLE >	ENDANGERED	CRITICALLY ENDANGERED	EXTINCT IN THE WILD	EXTINCT
NE	DD	LC	NT	VU	EN	CR	EW	EX

© David G. Long

Scaphophyllum speciosum is listed as 'Vulnerable' on the IUCN Red List of Threatened Species™. This large, beautiful liverwort is found in China (Tibet and Yunnan), Taiwan, Bhutan and East Nepal.

This species inhabits old growth mountain forest with high air humidity. Its habitat is extremely sensitive to disturbance, particularly to thinning and grazing by cattle, which opens up the forest to wind and sunshine, lowering humidity levels. The trees of these forests are highly valued for their timber, and so logging is a significant threat, further opening up the forest.

Scaphophyllum speciosum is now restricted to small, declining populations in fewer than ten locations.

Scaphophyllum speciosum is a large and conspicuous species, and is considered a good flagship species for highlighting the conservation status of bryophytes. In Bhutan, its population falls within the Thrumsengla National Park, but present levels of protection elsewhere are uncertain, and greater conservation efforts may be needed if the species is to survive in the long-term.

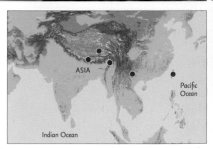

A species is Vulnerable when it faces a high risk of extinction in the wild, based on measurements of population size and/or geographic range and their trends in the past, present and/or future.

Seychelles Giant Millipede

Seychelleptus seychellarum

NOT EVALUATED	DATA DEFICIENT	LEAST CONCERN	NEAR THREATENED	< VULNERABLE >	ENDANGERED	CRITICALLY ENDANGERED	EXTINCT IN THE WILD	EXTINCT
NE	DD	LC	NT	VU	EN	CR	EW	EX

© Peter Hitchins

The **Seychelles Giant Millipede**, *Seychelleptus seychellarum*, is classified as 'Vulnerable' on the IUCN Red List of Threatened Species™. It is one of the largest millipedes in the world, regularly reaching over 15 cm in length. It only occurs on the Seychelles Islands of Cousin, Cousine, La Digue, Silhouette, Félicité, Aride and Frégate.

On most of these islands the Seychelles Giant Millipede is scarce, but there are concerns that it might be declining on the tiny 27 ha island of Cousine. It is very vulnerable to predation by rats, and it is also susceptible to being squashed by human traffic at night when the millipede wanders onto footpaths.

Conservation of the Seychelles Giant Millipede involves maintaining the forest canopy under which it lives. On Cousine Island, the forest canopy has been expanded by planting indigenous trees which has benefited its millipede population. Great care is also taken to avoid crushing this species along well-defined footpaths, which are only used with the help of a torch. Fortunately, Cousine Island is one of the very few tropical islands which is entirely free of rats.

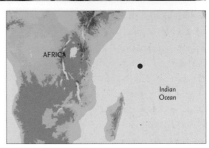

A species is Vulnerable when it faces a high risk of extinction in the wild, based on measurements of population size and/or geographic range and their trends in the past, present and/or future.

Hispaniolan Solenodon

Solenodon paradoxus

The **Hispaniolan Solenodon**, Solenodon paradoxus, is listed as 'Endangered' on the IUCN Red List of Threatened Species™. It is a large shrew-like animal and one of very few mammals capable of producing toxic saliva, which it injects into its invertebrate prey through grooves in its incisors. Mainly found in the Dominican Republic, it also clings on in Haiti's Massif de la Hotte, and occupies a range of habitats from coastal dry scrub up to high-elevation pine forest.

The Hispaniolan Solenodon represents a remarkable amount of unique evolutionary history, diverging from other living mammal groups some 75 million years ago, before the extinction of the dinosaurs. It is now threatened by widespread forest destruction through charcoal production and clearance for agriculture or urban development. It is also under pressure from predation by feral dogs and other invasive species.

The Dominican Republic's network of protected areas offers some protection to the Hispaniolan Solenodon's remaining habitat. A recently initiated project is conducting field research to assess the species' status, threats and specific conservation requirements.

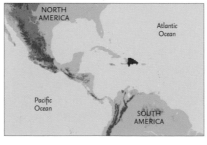

A species is Endangered when it faces a very high risk of extinction in the wild, based on measurements of population size and/or geographic range and their trends in the past, present and/or future.

Bulgarian Emerald

Somatochlora borisi

NOT EVALUATED	DATA DEFICIENT	LEAST CONCERN	NEAR THREATENED	⟨ VULNERABLE ⟩	ENDANGERED	CRITICALLY ENDANGERED	EXTINCT IN THE WILD	EXTINCT
NE	DD	LC	NT	VU	EN	CR	EW	EX

© Vincent Kalkman

The **Bulgarian Emerald**, *Somatochlora borisi*, is classified as 'Vulnerable' on the IUCN Red List of Threatened Species™. Compared to other groups of insects, dragonflies are relatively well-known, so it was therefore a surprise when this previously unknown species was discovered in Europe in 2001.

The Bulgarian Emerald is confined to large brooks and small rivers in the hills of southeast Bulgaria, eastern Greece and European Turkey. This beautiful area is still largely undisturbed and is mainly used for extensive wood production and the traditional rearing of goats and sheep. However, the development of *Robinia pseudacacia* plantations may reduce the availability of maturation and foraging areas for the species. Recent prolonged spells of hot and dry weather have also led to the desiccation of brooks and rivers. The expected increased impact of climate change is likely to put further pressure on this species.

Necessary conservation actions include the control of water pollution, removal of conifer plantations and conservation of clear riparian forests. Two rivers inhabited by this species are included in the Greek national Dadia Protected Area buffer zone.

A species is Vulnerable when it faces a high risk of extinction in the wild, based on measurements of population size and/or geographic range and their trends in the past, present and/or future.

Scalloped Hammerhead

NOT EVALUATED	DATA DEFICIENT	LEAST CONCERN	NEAR THREATENED	VULNERABLE	‹ ENDANGERED ›	CRITICALLY ENDANGERED	EXTINCT IN THE WILD	EXTINCT
NE	DD	LC	NT	VU	EN	CR	EW	EX

The **Scalloped Hammerhead**, *Sphyrna lewini*, is listed as 'Endangered' on the IUCN Red List of Threatened Species™. Named for the 'scalloped' front edge of its hammer-shaped head, this large shark is found worldwide in warm temperate and tropical waters.

Adult and juvenile Scalloped Hammerheads are taken as both a target species and by-catch in a range of fisheries, with the species' habit of coming together in large schools making it particularly vulnerable to capture, and making it appear more abundant than it actually is. The fins of hammerhead sharks are highly valued, and exploitation in fisheries is largely unregulated. The life history of the Scalloped Hammerhead gives it relatively low resilience to exploitation, and significant population declines have been reported.

Management plans, monitoring programmes and fishing regulations are urgently needed throughout this shark's range. The adoption of shark finning bans by some states, regions and fisheries organizations is accelerating, and it is hoped that this will increasingly prevent the capture of this and other shark species solely for their fins.

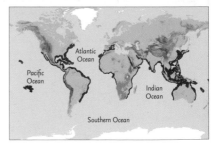

A species is Endangered when it faces a very high risk of extinction in the wild, based on measurements of population size and/or geographic range and their trends in the past, present and/or future.

Great Hammerhead

NOT EVALUATED	DATA DEFICIENT	LEAST CONCERN	NEAR THREATENED	VULNERABLE	⟨ ENDANGERED ⟩	CRITICALLY ENDANGERED	EXTINCT IN THE WILD	EXTINCT
NE	DD	LC	NT	VU	EN	CR	EW	EX

© JEREMY STAFFORD-DEITSCH

The **Great Hammerhead**, Sphyrna mokarran, is listed as 'Endangered' on the IUCN Red List of Threatened Species™. This large shark ranges widely throughout the tropical waters of the world, including the Atlantic, Pacific and Indian Oceans, and many smaller seas.

The Great Hammerhead is highly prized for its fins, and suffers very high levels of incidental mortality in other fisheries for tuna and tuna-like fishes. Like most other sharks, its slow growth and low reproduction rate makes it highly vulnerable to overexploitation. As a result, it has suffered serious declines, especially in parts of the eastern Atlantic where fishing effort is unmanaged and unmonitored.

The Great Hammerhead is in urgent need of coordinated conservation efforts involving the management of target and non-target fisheries. Fortunately, the increasing recognition of the detrimental effects of shark finning has led to the implementation of finning bans by fishing states in the United States, Australia and the European Union. Nevertheless, improved enforcement of legislation is required to prevent ongoing illegal finning activities.

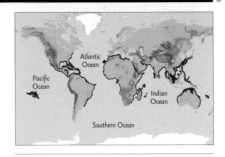

A species is Endangered when it faces a very high risk of extinction in the wild, based on measurements of population size and/or geographic range and their trends in the past, present and/or future.

NOT EVALUATED	DATA DEFICIENT	LEAST CONCERN	NEAR THREATENED	VULNERABLE	ENDANGERED	CRITICALLY ENDANGERED	EXTINCT IN THE WILD	EXTINCT
NE	DD	LC	NT	VU	EN	CR	EW	EX

Spruceanthus theobromae is listed as 'Critically Endangered' on the IUCN Red List of Threatened Species™. This rare liverwort grows on the bark of old cacao trees in rainforest at the foot of the Andes in the province of Los Rios in coastal Ecuador.

Since the 1960s, most of the region has been deforested, and no occurrences of the species in its natural habitat are now known. The species was long considered lost, but was rediscovered in 1997 in an old cacao plantation. An intensive field search for S. theobromae in 1999 revealed several additional localities, all in plantations with low management intensity and with little or no removal of epiphytes ('limpia') from the trunks of the cacao trees.

Intensification of cacao plantation management leads to local extinction of S. theobromae. The habitat of the species must therefore be considered to be severely threatened. Conservation of Spruceanthus theobromae is of considerable importance, as it is the only representative of the Asiatic genus Spruceanthus in the New World.

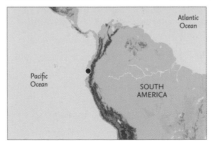

A species is Critically Endangered when it faces an extremely high risk of extinction in the wild, based on measurements of population size and/or geographic range and their trends in the past, present and/or future.

Spiny Dogfish

Squalus acanthias

RED LIST

NOT EVALUATED	DATA DEFICIENT	LEAST CONCERN	NEAR THREATENED	‹ VULNERABLE ›	ENDANGERED	CRITICALLY ENDANGERED	EXTINCT IN THE WILD	EXTINCT
NE	DD	LC	NT	VU	EN	CR	EW	EX

© ANDY MURCH

The **Spiny Dogfish**, *Squalus acanthias*, is listed as 'Vulnerable' on the IUCN Red List of Threatened Species™. This relatively small shark is principally found in coastal waters of the eastern and western Atlantic, the southern coasts of Australia and New Zealand, the eastern and western North Pacific, and the eastern South Pacific.

The Spiny Dogfish is targeted as a valuable commercial species and suffers high levels of mortality due to accidental by-catch in other fisheries. Although previously naturally abundant, this shark is vulnerable to overexploitation because of its late maturity, low reproductive capacity, and long generation time. It also has the longest pregnancy of any animal. Two subpopulations in the northwest and northeast Atlantic Ocean are considered to be at particularly high risk.

Despite several decades of warnings of unsustainable fishing pressure and reported steep stock declines, very few conservation or management measures are in place for the Spiny Dogfish, while measures that are in place have been relatively ineffective. Perhaps the only exception is in New Zealand, where quotas have been introduced to limit catches to sustainable levels.

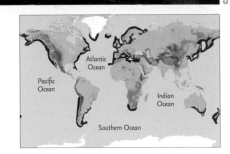

A species is Vulnerable when it faces a high risk of extinction in the wild, based on measurements of population size and/or geographic range and their trends in the past, present and/or future.

Angel Shark

Squatina squatina

NOT EVALUATED	DATA DEFICIENT	LEAST CONCERN	NEAR THREATENED	VULNERABLE	ENDANGERED	CRITICALLY ENDANGERED	EXTINCT IN THE WILD	EXTINCT
NE	DD	LC	NT	VU	EN	CR	EW	EX

The **Angel Shark**, *Squatina squatina*, is listed as 'Critically Endangered' on the IUCN Red List of Threatened Species™. Historically, the Angel Shark's range extended from Scandinavia to northwestern Africa, including the Mediterranean and the Black Sea. However, it has now vanished from some areas, and is extremely uncommon throughout most of the remainder of its range.

Although not particularly sought after by fisheries, the Angel Shark's habit of lying on the sea bottom makes it particularly vulnerable to by-catch in trawl fisheries.

Over the last 50 years, trawling activity has increased, and as a result the population has declined dramatically, and has even been declared extinct in the North Sea.

All *Squatina* species are protected within three Balearic Islands marine reserves, where fishing for these species is forbidden. There is an urgent need to confirm the status of the Angel Shark in the southern Mediterranean, Canary Islands and other areas where populations may still persist, so that appropriate conservation measures can be implemented as soon as possible.

A species is **Critically Endangered** when it faces an extremely high risk of extinction in the wild, based on measurements of population size and/or geographic range and their trends in the past, present and/or future.

Trapdoor Spider

Stasimopus robertsi

RED LIST

NOT EVALUATED	DATA DEFICIENT	LEAST CONCERN	NEAR THREATENED	VULNERABLE	‹ ENDANGERED ›	CRITICALLY ENDANGERED	EXTINCT IN THE WILD	EXTINCT
NE	DD	LC	NT	VU	EN	CR	EW	EX

© Les Oates

The **Trapdoor Spider**, *Stasimopus robertsi*, has not yet been officially assessed for the IUCN Red List of Threatened Species™, however it has a potential classification of 'Endangered'. It is endemic to South Africa and is found in the central part of the Gauteng Province.

The Trapdoor Spider is a nocturnal burrowing spider. Their burrows are closed off with hinged trapdoors of variable thickness, which are used for protection and as a prey detection system. Usually during the day, and during harsh weather, moulting and egg-laying, the lids are kept closed.

They open the trapdoor slightly, waiting for prey to pass close to the entrance, then rush out, grab the prey and return to the burrow.

The main threat to these spiders is the destruction of their habitat. The area where they were originally collected from is now mostly urban, with very little natural vegetation left. In the last few years the odd specimen has been found, but only in gardens. Further research into their ecology, taxonomy, and distribution has been recommended to help conserve this species.

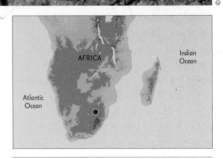

A species is Endangered when it faces a very high risk of extinction in the wild, based on measurements of population size and/or geographic range and their trends in the past, present and/or future.

Chinese Crested Tern

Sterna bernsteini

NOT EVALUATED	DATA DEFICIENT	LEAST CONCERN	NEAR THREATENED	VULNERABLE	ENDANGERED	CRITICALLY ENDANGERED	EXTINCT IN THE WILD	EXTINCT
NE	DD	LC	NT	VU	EN	CR	EW	EX

The **Chinese Crested Tern**, *Sterna bernsteini*, is listed as 'Critically Endangered' on the IUCN Red List of Threatened Species™. This seabird was previously feared to be extinct but was rediscovered in 2000 when, for the first time, its breeding grounds were located. Exclusively coastal, the Chinese Crested Tern has recently been recorded breeding at only two sites in China, Zhejiang and Fujian Provinces, and outside the breeding season on Halmahera, Indonesia, in Sarawak, Malaysia, Taiwan (China), Thailand and the Philippines.

The total population is estimated to be fewer than 50 individuals. The future of this species is threatened by egg-collection (it nests in colonies of other seabirds and collectors do not realise how rare it is) and human disturbance. Other threats include the effects of predatory species such as rats, and the risk of water pollution that affect the terns' food sources.

The Chinese Crested Tern is nationally protected in China and Thailand, and the nesting habitat is patrolled in the breeding season. This has served as an effective deterrent to egg collectors. Programmes to raise local awareness of this bird, and address socio-economic factors relating to its conservation, are being developed.

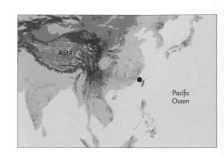

Smooth Cauliflower Coral

Stylophora pistillata

NOT EVALUATED	DATA DEFICIENT	LEAST CONCERN	NEAR THREATENED	VULNERABLE	ENDANGERED	CRITICALLY ENDANGERED	EXTINCT IN THE WILD	EXTINCT
NE	DD	LC	NT	VU	EN	CR	EW	EX

Smooth Cauliflower Coral, *Stylophora pistillata*, is listed as 'Near Threatened' on the IUCN Red List of Threatened Species™. Sometimes referred to as the Tramp Coral, due to the tendency of the larvae to attach to floating objects on which they travel hundreds of kilometres, this species ranges widely in the Indo-West Pacific.

The principal threat to the world's corals is the rise in sea temperature associated with global climate change. This leads to coral bleaching, where the symbiotic algae are expelled, leaving the corals weak and vulnerable. Climate change is also expected to increase ocean acidification and result in a greater frequency of extreme weather events. This is in addition to the localised threats from the aquarium trade, pollution, destructive fishing practices, invasive species, human development, and other activities.

Smooth Cauliflower Coral is listed on Appendix II of CITES, which means that trade in this species should be carefully regulated. It also occurs in several Marine Protected Areas across its range. One of the main priorities now is to conduct further research into its ecology, population biology, threats and recovery options.

© David Obura

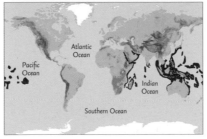

A species is Near Threatened when it does not qualify for Critically Endangered, Endangered or Vulnerable now, but is close to qualifying for or is likely to qualify for a threatened category in the near future.

NOT EVALUATED	DATA DEFICIENT	LEAST CONCERN	NEAR THREATENED	⟨ VULNERABLE ⟩	ENDANGERED	CRITICALLY ENDANGERED	EXTINCT IN THE WILD	EXTINCT
NE	DD	LC	NT	VU	EN	CR	EW	EX

The **Big-leaf Mahogany**, *Swietenia macrophylla*, is listed as 'Vulnerable' on the IUCN Red List of Threatened Species™. This towering tree has a patchy distribution from southern Mexico, south through Central America to Brazil and Bolivia.

The beautiful hardwood obtained from mahogany trees has been in high demand for centuries. Following the commercial extinction of Caribbean Mahogany and Honduras Mahogany due to overexploitation, the Big-leaf Mahogany has become the most commercially important member of the genus. However, many people in the industry are concerned that without enforced protection measures, this species may also become commercially extinct in the near future.

In a vital move for its protection, Big-leaf Mahogany was included on Appendix II of the Convention on International Trade in Endangered Species (CITES) in 2002. Appendix II includes species not necessarily threatened with extinction but for which trade must be controlled in order to avoid threats to their survival. It is hoped that this legislation will enable the sustainable management of this important and majestic tree.

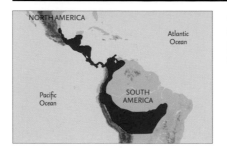

A species is **Vulnerable** when it faces a high risk of extinction in the wild, based on measurements of population size and/or geographic range and their trends in the past, present and/or future.

Gilded Presba

Syncordulia legator

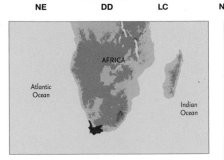

A species is Vulnerable when it faces a high risk of extinction in the wild, based on measurements of population size and/or geographic range and their trends in the past, present and/or future.

© MICHAEL SAMWAYS

The **Gilded Presba**, Syncordulia legator, is classified as 'Vulnerable' on the IUCN Red List of Threatened Species™. Only scientifically described in 2007, it is a large dragonfly that lives in the Cape Fold mountains of South Africa. It is capable of remarkably powerful flight, emerging from the cool mountain streams and then moving swiftly up and over the mountains, even in windy conditions.

The main threat to the Gilded Presba is from invasive alien trees which aggressively spread along the stream and river banks, shading out its habitat. These alien trees eventually squeeze out much of the local flora and fauna.

The nationwide Working for Water Programme in South Africa is removing alien trees along water courses. This activity is conferring a huge and immediate benefit upon many species. Without such a programme, which was not originally planned specifically for species survival, it is quite likely that in years to come the Gilded Presba may well have gone extinct without us ever having known of its existence.

NOT EVALUATED	DATA DEFICIENT	LEAST CONCERN	NEAR THREATENED	< VULNERABLE >	ENDANGERED	CRITICALLY ENDANGERED	EXTINCT IN THE WILD	EXTINCT
NE	DD	LC	NT	VU	EN	CR	EW	EX

Takakia ceratophylla is listed as 'Vulnerable' on the IUCN Red List of Threatened Species™. This moss represents a primitive group of bryophytes, and has an important position in the study and understanding of the evolution of land plants. It is found in just a few widely separated locations in the USA (Aleutian Islands), Sikkim, Nepal and China (Tibet, Yunnan).

Takakia ceratophylla grows mainly on shaded, damp cliffs, sheltered rock faces and very wet ground with late snow cover, sometimes forming luxuriant greenish cover in suitable habitat. However, in all localities its populations are very small, and are under threat from habitat disturbance caused by human activities. Climate change also poses a potentially serious threat, and the species' restriction to just a few small, fragmented areas makes it particularly vulnerable.

There are not known to be any specific conservation measures currently in place for this moss. It has much to reveal about bryophyte evolution, but its scattered populations will need full protection if the species is to survive long-term.

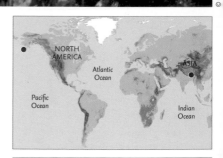

A species is Vulnerable when it faces a high risk of extinction in the wild, based on measurements of population size and/or geographic range and their trends in the past, present and/or future.

Malayan Tapir

Tapirus indicus

NOT EVALUATED	DATA DEFICIENT	LEAST CONCERN	NEAR THREATENED	VULNERABLE	‹ ENDANGERED ›	CRITICALLY ENDANGERED	EXTINCT IN THE WILD	EXTINCT
NE	DD	LC	NT	VU	EN	CR	EW	EX

© Anders Gonçalves da Silva

The **Malayan Tapir**, *Tapirus indicus*, is listed as 'Endangered' on the IUCN Red List of Threatened Species™. It has the distinction of being the largest of the four tapir species, as well as being the only tapir native to the Old World. The Malayan Tapir is unmistakable with its bold black and white markings.

Once widely abundant, over recent decades Malayan Tapir population numbers have rapidly declined, and the species now survives only as isolated populations in remote or protected areas in Indonesia, Myanmar, Peninsular Malaysia, and Thailand. Habitat destruction poses the predominant threat, as a result of forests being cleared for human settlement, agriculture and, more recently, palm oil plantations. This species is also hunted for its meat and for sale in the Asian zoo trade, and often becomes road-kill.

International trade in the Malayan Tapir is prohibited under its listing on Appendix I of CITES. It is also legally protected in all countries in which it occurs, and is found in a number of protected areas, including some of the most secure reserves in Southeast Asia.

A species is **Endangered** when it faces a very high risk of extinction in the wild, based on measurements of population size and/or geographic range and their trends in the past, present and/or future.

NOT EVALUATED	DATA DEFICIENT	LEAST CONCERN	NEAR THREATENED	VULNERABLE	**‹ ENDANGERED ›**	CRITICALLY ENDANGERED	EXTINCT IN THE WILD	EXTINCT
NE	DD	LC	NT	VU	EN	CR	EW	EX

© DIEGO L LIZCANO

The **Mountain Tapir**, Tapirus pinchaque, is listed as 'Endangered' on the IUCN Red List of Threatened Species™. This species survives in a few remaining undisturbed refuges high in the Andes of Colombia, Ecuador and Peru.

The major threats to the Mountain Tapir are destruction and fragmentation of its habitat, illegal hunting for food, and the use of body parts in folk medicine. The introduction of livestock to the area has also introduced new diseases and attracted more predators into its range.

The species is protected by law throughout its distribution and is listed on an international convention (CITES) to ensure trade does not threaten its survival. However, illegal hunting remains a major threat. Populations occur within a number of protected reserves in Colombia and Ecuador and a small number of individuals are held in zoos. The IUCN/ SSC Tapir Specialist Group (TSG) was created in 1980 to promote the research and conservation of the four species of tapirs in their areas of occurrence in South and Central America, and Southeast Asia.

A species is Endangered when it faces a very high risk of extinction in the wild, based on measurements of population size and/or geographic range and their trends in the past, present and/or future.

Titicaca Water Frog

Telmatobius culeus

NOT EVALUATED	DATA DEFICIENT	LEAST CONCERN	NEAR THREATENED	VULNERABLE	ENDANGERED	CRITICALLY ENDANGERED	EXTINCT IN THE WILD	EXTINCT
NE	DD	LC	NT	VU	EN	CR	EW	EX

© Ignacio De La Riva

The **Titicaca Water Frog**, *Telmatobius culeus*, is listed as 'Critically Endangered' on the IUCN Red List of Threatened Species™. It is found in Lake Titicaca in Peru and Bolivia. It is the largest aquatic frog, spending its entire life underwater (it is thought to breathe through its loose-fitting skin).

The number of Titicaca Water Frogs has declined dramatically in recent years as a result of harvesting for human consumption, degradation of its habitat, and the extraction of water from the lake.

The predation of larvae by introduced trout species is also thought to be a problem.

Habitat management and protection is present at the Lake Titicaca Reserve, but further action and better enforcement are desperately needed. Whilst captive-breeding programmes have taken place, they have so far been unsuccessful.

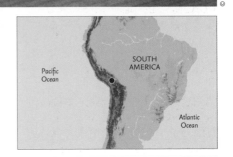

A species is Critically Endangered when it faces an extremely high risk of extinction in the wild, based on measurements of population size and/or geographic range and their trends in the past, present and/or future.

Terpsiphone corvina

NOT EVALUATED	DATA DEFICIENT	LEAST CONCERN	NEAR THREATENED	VULNERABLE	ENDANGERED	CRITICALLY ENDANGERED	EXTINCT IN THE WILD	EXTINCT
NE	DD	LC	NT	VU	EN	CR	EW	EX

RED LIST

A species is Critically Endangered when it faces an extremely high risk of extinction in the wild, based on measurements of population size and/or geographic range and their trends in the past, present and/or future.

The **Seychelles Paradise-flycatcher**, *Terpsiphone corvina*, is listed as 'Critically Endangered' on the IUCN Red List of Threatened Species™. This elegant flycatcher is endemic to the Seychelles, where it is known in Creole as 'Veuve', from the French for 'widow', owing to its all black plumage.

The only viable population of the Seychelles Paradise-flycatcher is now restricted to La Digue. Habitat loss, due to tourism and private housing developments, has been the major cause of the precipitous decline in the population, and continues to be the greatest threat to this species. Recently, a disease affecting Takamaka trees, leading to increased woodland clearance, has provided further cause for concern.

Owing to various management measures, including the designation of a small reserve and a public awareness programme, the Seychelles Paradise-flycatcher population has shown a small increase in recent decades. In order to expand this species' range and thus reduce its vulnerability, translocations to other islands in the archipelago are under way, with 23 birds reintroduced to the now restored Denis in 2008, where they bred the following year.

Egyptian Tortoise

Testudo kleinmanni

NOT EVALUATED	DATA DEFICIENT	LEAST CONCERN	NEAR THREATENED	VULNERABLE	ENDANGERED	CRITICALLY ENDANGERED	EXTINCT IN THE WILD	EXTINCT
NE	DD	LC	NT	VU	EN	CR	EW	EX

The **Egyptian Tortoise**, *Testudo kleinmanni*, is listed as 'Critically Endangered' on the IUCN Red List of Threatened Species™. This Middle Eastern and North African tortoise has a golden-coloured shell and minute shell size of less than 14 cm.

Currently, the main threats endangering this species are intensive commercial collection for the national and international pet trade, and habitat destruction which has led to its disappearance from much of its former range in scrub desert and coastal dunes. Agricultural expansion, cultivation, overgrazing, and urban encroachment have put enormous pressure on the Egyptian Tortoise's fragile and dwindling habitat, dramatically reducing available vegetation for food and cover.

Nationally protected in Egypt and Israel, including through community-based protected areas, but not in Libya, this species is also protected from international trade, although these laws are often not enforced. Its future will depend on the establishment of more protected areas, wider engagement with local communities, and improved enforcement of trade laws.

© Omar Attum

A species is **Critically Endangered** when it faces an extremely high risk of extinction in the wild, based on measurements of population size and/or geographic range and their trends in the past, present and/or future.

Giant Garter Snake

NOT EVALUATED	DATA DEFICIENT	LEAST CONCERN	NEAR THREATENED	‹ VULNERABLE ›	ENDANGERED	CRITICALLY ENDANGERED	EXTINCT IN THE WILD	EXTINCT
NE	DD	LC	NT	VU	EN	CR	EW	EX

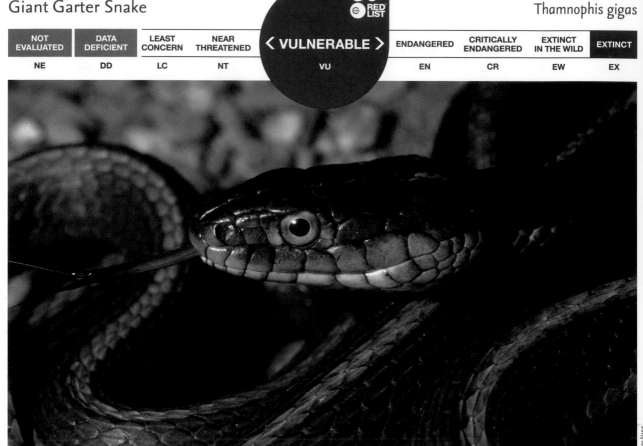

The **Giant Garter Snake**, Thamnophis gigas, is listed as 'Vulnerable' on the IUCN Red List of Threatened Species™. The largest of the garter snake species, this highly aquatic snake occurs in the Central Valley of California, in the western United States.

The Giant Garter Snake has been lost from much of its former range due to the loss, degradation and fragmentation of its wetland habitats. Flood control, pollution, changes in land use and agricultural practices, overgrazing, and contamination with heavy metals all threaten the species and its habitat, and individuals may also fall prey to introduced predators such as cats, bullfrogs and large predatory fish.

A recovery plan for the species recommended a number of conservation actions, including the protection of existing populations and habitat, restoration of former habitat, further surveys, research and monitoring, and outreach and incentive programmes for the public. The Giant Garter Snake occurs in a number of wildlife refuges.

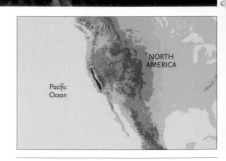

A species is Vulnerable when it faces a high risk of extinction in the wild, based on measurements of population size and/or geographic range and their trends in the past, present and/or future.

NOT EVALUATED	DATA DEFICIENT	LEAST CONCERN	NEAR THREATENED	VULNERABLE	ENDANGERED	CRITICALLY ENDANGERED	EXTINCT IN THE WILD	EXTINCT
NE	DD	LC	NT	VU	EN	CR	EW	EX

A species is **Critically Endangered** when it faces an extremely high risk of extinction in the wild, based on measurements of population size and/or geographic range and their trends in the past, present and/or future.

© Bedgebury Pinetum

Thuja sutchuenensis is listed as 'Critically Endangered' on the IUCN Red List of Threatened Species™. Formerly classified as 'Extinct in the Wild', this rare conifer was rediscovered in 1999, not far from the locality in central China where the only known specimens were collected a century before.

Owing to the suitability of its wood for construction and craftwork, many of the most accessible trees have been cut down, with the remaining population estimated at just 200 individuals. It is probable that the species has gone through a genetic bottleneck, and thus inbreeding amongst the remaining individuals may be a serious problem. In addition, seedlings are scarce, and plant diseases caused by microorganisms have been observed.

As the only member of its genus in central China, Thuja sutchuenensis is considered to be of great scientific value and a high conservation priority. The primary objectives of an ongoing conservation project are to conduct further surveys of potential habitat, to identify the causes of the species' poor regeneration, and to propagate seedlings in nurseries.

Delacour's Langur

NOT EVALUATED	DATA DEFICIENT	LEAST CONCERN	NEAR THREATENED	VULNERABLE	ENDANGERED	CRITICALLY ENDANGERED	EXTINCT IN THE WILD	EXTINCT
NE	DD	LC	NT	VU	EN	CR	EW	EX

© Terry Whittaker

Delacour's Langur, *Trachypithecus delacouri*, is listed as 'Critically Endangered' on the IUCN Red List of Threatened Species™. Found only in north-central Vietnam, Delacour's Langur is one of the rarest and most threatened primates on Earth.

With a population estimated to be less than 250 mature individuals, fragmented into numerous subpopulations, most of which are unviable in the long-term, this species is dangerously close to extinction. The level of hunting has reduced in recent years as a result of increased legal protection, but continues to pose a significant threat, whilst ongoing habitat loss is contributing to the further isolation of the remaining populations.

Recent commitments by the Vietnamese Government and several international conservation organisations to do more to protect Delacour's Langur have greatly improved the chances of saving this rare and remarkable species. The main priority is to increase the level of protection and law enforcement within those protected areas that support the remaining populations.

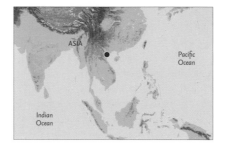

ASIA

Pacific Ocean

Indian Ocean

A species is **Critically Endangered** when it faces an extremely high risk of extinction in the wild, based on measurements of population size and/or geographic range and their trends in the past, present and/or future.

Andean Bear

Tremarctos ornatus

© ROBYN APPLETON

NOT EVALUATED	DATA DEFICIENT	LEAST CONCERN	NEAR THREATENED	‹ VULNERABLE ›	ENDANGERED	CRITICALLY ENDANGERED	EXTINCT IN THE WILD	EXTINCT
NE	DD	LC	NT	VU	EN	CR	EW	EX

RED LIST

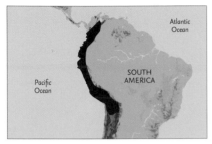

A species is **Vulnerable** when it faces a high risk of extinction in the wild, based on measurements of population size and/or geographic range and their trends in the past, present and/or future.

The **Andean Bear**, Tremarctos ornatus, is listed as 'Vulnerable' on the IUCN Red List of Threatened Species™. The only bear species in South America, it occurs in the Andes Mountains, from western Venezuela to southern Bolivia, occupying habitats ranging from desert-scrub to forests and high altitude grasslands. Most individuals have white, spectacle-like markings around the eyes, giving rise to the alternative name 'Spectacled Bear'.

The main threat to the Andean Bear is from habitat loss as a result of agricultural expansion, grazing, mining, oil exploration, and road development. Due to the reduction and fragmentation of their habitat, Andean Bears increasingly raid crops and kill livestock, resulting in more retaliatory killing and illegal hunting.

The Andean Bear occurs in a number of protected areas, but many do not provide adequate protection or are too small and isolated to support viable bear populations. Although it is protected by national laws in all five range states, and through its listing on Appendix I of CITES, stricter enforcement is needed. Further research is required to better manage human-bear conflict.

West African Manatee

Trichechus senegalensis

A species is Vulnerable when it faces a high risk of extinction in the wild, based on measurements of population size and/or geographic range and their trends in the past, present and/or future.

© TOMAS DIAGNE, OCEANIUM DAKAR

The **West African Manatee**, *Trichechus senegalensis*, is listed as 'Vulnerable' on the IUCN Red List of Threatened Species™. The least-known of the Sirenians (Manatees and Dugongs), the West African Manatee inhabits shallow coastal waters, wetland systems and rivers from Senegal to Angola.

Incidental capture in nets, and hunting for meat, skin, bones and oil, occurs throughout most of its range. Habitat loss is an additional threat, as a result of the damming of rivers, cutting of mangroves for firewood, coastal development, and the destruction of wetlands for agricultural development. The situation is likely to be exacerbated by climate change.

Although the West African Manatee is listed on Appendix II of CITES and is protected by national laws in all countries where it lives, high levels of human poverty in the region make effective conservation extremely challenging. Fortunately, fifteen countries and three NGOs have recently agreed to work together to conserve the manatees and small cetaceans of West Africa and the Macaronesian Islands, based on a Manatee Action Plan.

NOT EVALUATED	DATA DEFICIENT	LEAST CONCERN	NEAR THREATENED	VULNERABLE	⟨ ENDANGERED ⟩	CRITICALLY ENDANGERED	EXTINCT IN THE WILD	EXTINCT
NE	DD	LC	NT	VU	EN	CR	EW	EX

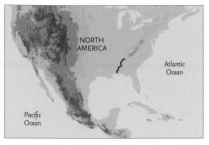

A species is Endangered when it faces a very high risk of extinction in the wild, based on measurements of population size and/or geographic range and their trends in the past, present and/or future.

© ARTHUR E. BOGAN, NORTH CAROLINA STATE MUSEUM OF NATURAL SCIENCES

The **Tulotoma Snail**, Tulotoma magnifica, is listed as 'Endangered' on the IUCN Red List of Threatened Species™. This rare freshwater snail is endemic to the Coosa and Alabama River systems in Alabama, USA.

The Tulotoma Snail has suffered major declines due to a variety of human influences, including dredging, dam construction, siltation, hydropower discharge, and pollution. As a result, this species was listed on the US Endangered Species List in 1991. Fortunately, later surveys revealed that its distribution is much greater than once thought, and all the known populations are either stable or increasing.

Researchers are continuing to survey additional sites within the historical range of the Tulotoma Snail, but current data indicates that this species no longer warrants classification as an Endangered species. However, given that the surviving populations are still isolated and vulnerable to pollution, as well as to floods and droughts, the classification of the Tulotoma Snail as 'Endangered' is still deemed to be appropriate.

NOT EVALUATED	DATA DEFICIENT	LEAST CONCERN	NEAR THREATENED	VULNERABLE	ENDANGERED	CRITICALLY ENDANGERED	EXTINCT IN THE WILD	EXTINCT
NE	DD	LC	NT	VU	EN	CR	EW	EX

Turbinicarpus alonsoi is listed as 'Critically Endangered' on the IUCN Red List of Threatened Species™. This small, attractive cactus occurs in a single location of less than 10 km² in the State of Querétaro, Mexico, and has a current estimated population size of less than 5,000 individuals.

This species' habitat has become widely known to collectors, and since its discovery in 1996, the population has declined by more than 50 percent due to illegal over-collecting. Its highly restricted range

also puts the cactus at increased risk of being wiped out by any chance events.

Although Turbinicarpus alonsoi is listed on Appendix I of CITES, which prohibits all international trade in the species, the laws governing imports in the countries of destination need to be much more strongly enforced. This rare cactus is currently propagated in Europe from illegally collected plants.

A species is **Critically Endangered** when it faces an extremely high risk of extinction in the wild, based on measurements of population size and/or geographic range and their trends in the past, present and/or future.

NOT EVALUATED	DATA DEFICIENT	LEAST CONCERN	NEAR THREATENED	VULNERABLE	ENDANGERED	CRITICALLY ENDANGERED	EXTINCT IN THE WILD	EXTINCT
NE	DD	LC	NT	VU	EN	CR	EW	EX

Uebelmannia buiningii is listed as 'Critically Endangered' on the IUCN Red List of Threatened Species™. With its population fragmented into numerous tiny subpopulations in the Serra Negra in the Brazilian state of Minas Gerais, this diminutive cactus is believed to be on the verge of extinction.

Like many cacti, one of the main threats facing U. buiningii is the illegal collection of wild plants and seeds for the horticultural trade. In addition to harvesting by collectors, this species is also affected by fires and trampling by cattle.

There are no specific conservation measures in place for U. buiningii and it is not thought to be common in cultivation. Nonetheless, it is listed on Appendix I of CITES, meaning that all international trade in this species is prohibited.

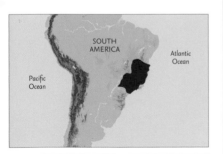

A species is **Critically Endangered** when it faces an extremely high risk of extinction in the wild, based on measurements of population size and/or geographic range and their trends in the past, present and/or future.

© Graham Charles

Polar Bear

Ursus maritimus

NOT EVALUATED	DATA DEFICIENT	LEAST CONCERN	NEAR THREATENED	⟨ VULNERABLE ⟩	ENDANGERED	CRITICALLY ENDANGERED	EXTINCT IN THE WILD	EXTINCT
NE	DD	LC	NT	VU	EN	CR	EW	EX

The **Polar Bear**, *Ursus maritimus*, is classified as 'Vulnerable' on the IUCN Red List of Threatened Species™. The polar bear is the largest living land carnivore in the world today and lives throughout the ice-covered waters of the circumpolar Arctic.

It is recognised that climate change has already had an impact on some polar bears and their sea-ice habitat, affecting access to their prey and to den areas. Persistent organic pollutants (POPs) also pose a threat to polar bears – if accumulated at elevated levels, these compounds can cause neurological, reproductive and immunological changes. Hunting of polar bear is now controlled, although in some places overharvesting is a concern that is being addressed.

A number of countries have signed the 'International Agreement on the Conservation of Polar Bears', which identifies the right of local hunters to harvest polar bears sustainably, and to outlaw hunting from aircraft and large ships. The threats caused by climate change are now the main concern – the complexity of these issues will demand international cooperation if this species is to survive.

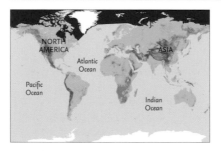

A species is Vulnerable when it faces a high risk of extinction in the wild, based on measurements of population size and/or geographic range and their trends in the past, present and/or future.

NOT EVALUATED	DATA DEFICIENT	LEAST CONCERN	NEAR THREATENED	⟨ VULNERABLE ⟩	ENDANGERED	CRITICALLY ENDANGERED	EXTINCT IN THE WILD	EXTINCT
NE	DD	LC	NT	VU	EN	CR	EW	EX

© Dave Garshelis

The **Asiatic Black Bear**, ('Moon Bear'), *Ursus thibetanus*, is listed as 'Vulnerable' on the IUCN Red List of Threatened Species™. It has a patchy range roughly coinciding with forest distribution from southeastern Iran, through southern and eastern Asia, and northwards to Japan, Korea, and the Russian Far East.

This species has been hunted for millennia for its skin, paws and gall bladder, from which bile is obtained for use in traditional Chinese medicine. The illegal killing of bears, combined with extensive habitat loss, is causing rapid and widespread population declines, especially in Southeast Asia and China. The commercial farming of thousands of bears in China and Vietnam for bile extraction, ostensibly to save wild bears, has no demonstrated conservation value.

Although Asiatic Black Bears are legally protected in most of the 18 range countries, poaching (especially by snaring and poisoning) is rampant. These bears are also killed when ransacking crops. The control of poaching and smuggling, and the protection of forest habitats are important conservation priorities, vital to the species' survival.

A species is Vulnerable when it faces a high risk of extinction in the wild, based on measurements of population size and/or geographic range and their trends in the past, present and/or future.

NOT EVALUATED	DATA DEFICIENT	LEAST CONCERN	NEAR THREATENED	VULNERABLE	⟨ENDANGERED⟩	CRITICALLY ENDANGERED	EXTINCT IN THE WILD	EXTINCT
NE	DD	LC	NT	VU	EN	CR	EW	EX

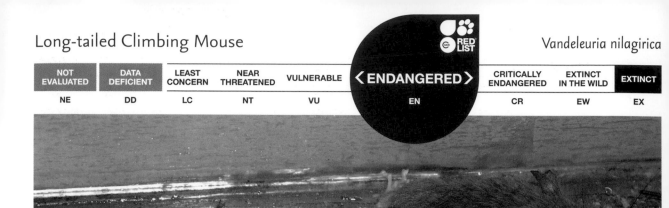

The **Long-tailed Climbing Mouse**, *Vandeleuria nilagirica*, is listed as 'Endangered' on the IUCN Red List of Threatened Species™. This small, arboreal murid is endemic to the northern Western Ghats of India, where it is found in evergreen montane forests and relatively undisturbed coffee, banana and cardamon plantations, at elevations of between 900 and 2,100 metres.

The Long-tailed Climbing Mouse is primarily threatened by a decline in the extent of its forest habitat and the replacement of native canopy trees with exotic species. Surviving in isolated forest patches, the distribution of this species is highly fragmented, with the only known sizeable population residing at Haleri in Coorg on the Western Ghats.

There are no known conservation measures in place for the Long-tailed Climbing Mouse. However, with so little known about the status of this species in the wild, further studies into its ecology, distribution and population will be required if its future is to be secured.

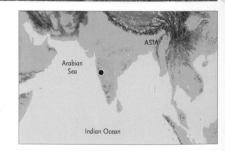

A species is Endangered when it faces a very high risk of extinction in the wild, based on measurements of population size and/or geographic range and their trends in the past, present and/or future.

Sociable Lapwing

Vanellus gregarius

RED LIST

NOT EVALUATED	DATA DEFICIENT	LEAST CONCERN	NEAR THREATENED	VULNERABLE	ENDANGERED	CRITICALLY ENDANGERED	EXTINCT IN THE WILD	EXTINCT
NE	DD	LC	NT	VU	EN	CR	EW	EX

© JAN-MICHAEL BREIDER, WWW.PBASE.COM/BRE DER

The **Sociable Lapwing**, *Vanellus gregarius*, is listed as 'Critically Endangered' on the IUCN Red List of Threatened Species™. It breeds in Russia and Kazakhstan and winters in Israel, Syria, Eritrea, Sudan and northwest India. It may also be found in Pakistan, The Gulf and Oman during the winter.

The Sociable Lapwing is suffering a severe range reduction and its numbers are dwindling. Reasons for this decline in numbers are poorly understood. The main problem appears to be loss of short-grass steppe following the reduction of Soviet-subsidised cattle farming in Kazakhstan, although hunting during migration is currently thought to pose the greatest threat.

This species has legal protection in Armenia, Kazakhstan, Russia, Turkmenistan, Ukraine and Uzbekistan, but this is generally not enforced. Better management of its grassland habitats, effective management of colonies during the nesting season, and protection against hunting in Syria (and elsewhere on its migration routes) will help to prevent the extinction of the Sociable Lapwing.

A species is *Critically Endangered* when it faces an extremely high risk of extinction in the wild, based on measurements of population size and/or geographic range and their trends in the past, present and/or future.

Komodo Dragon

Varanus komodoensis

NOT EVALUATED	DATA DEFICIENT	LEAST CONCERN	NEAR THREATENED	⟨ VULNERABLE ⟩	ENDANGERED	CRITICALLY ENDANGERED	EXTINCT IN THE WILD	EXTINCT
NE	DD	LC	NT	VU	EN	CR	EW	EX

The **Komodo Dragon**, *Varanus komodoensis*, is listed as 'Vulnerable' on the IUCN Red List of Threatened Species™. It is the largest living species of lizard, growing to an average length of 2 to 3 metres and weighing upwards of 70 kg, typically found in dry open grasslands, savannah and tropical forest habitats at low elevations.

It is estimated that there are between 3,000 and 5,000 Komodo Dragons left on Earth, although populations are heavily fragmented and spread between five small Indonesian islands. Habitat fragmentation significantly increases the risk posed by natural chance events, such as volcanoes, wildfires, tsunamis, etc. However, burning of grasslands by humans and subsequent predation on prey species such as Rusa Deer, *Rusa timorensis*, is probably the major threat to this species at present.

The Komodo National Park was established in 1980 to conserve the Komodo Dragon and the ecosystem in which it lives. Smaller reserves also help give some degree of protection, although, as in the Park, law enforcement is often lacking and human-dragon conflicts continue to exist.

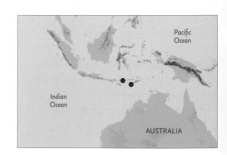

Panay Monitor Lizard

Varanus mabitang

NOT EVALUATED	DATA DEFICIENT	LEAST CONCERN	NEAR THREATENED	VULNERABLE	⟨ ENDANGERED ⟩	CRITICALLY ENDANGERED	EXTINCT IN THE WILD	EXTINCT
NE	DD	LC	NT	VU	EN	CR	EW	EX

© Tim Laman/www.TimLaman.com

The **Panay Monitor Lizard**, *Varanus mabitang*, is listed as 'Endangered' on the IUCN Red List of Threatened Species™. This recently described monitor lizard is endemic to the island of Panay in the Philippines. A highly arboreal and specialized frugivore, it is generally associated with large trees in primary tropical moist forest.

Given that only 12 individuals have been caught since 2002, the Panay Monitor Lizard appears to be naturally very rare. Furthermore, its forest habitat is under significant threat from the conversion of land for agricultural use and logging operations. In addition to the threat of habitat loss, this lizard is a favourite food animal, and overhunting represents a serious concern.

Fortunately, several non-governmental organizations are conducting ongoing projects on Panay to conserve remaining areas of suitable habitat, and to research the biology and habitat requirements of the Panay Monitor Lizard. It is also listed on Appendix II of CITES, which prohibits all international trade without a permit, and is present in the Central Panay and Northwest Panay protected areas.

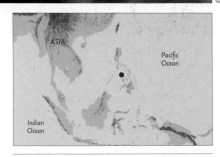

A species is Endangered when it faces a very high risk of extinction in the wild, based on measurements of population size and/or geographic range and their trends in the past, present and/or future.

Forest Coconut

NOT EVALUATED	DATA DEFICIENT	LEAST CONCERN	NEAR THREATENED	VULNERABLE	ENDANGERED	CRITICALLY ENDANGERED	EXTINCT IN THE WILD	EXTINCT
NE	DD	LC	NT	VU	EN	CR	EW	EX

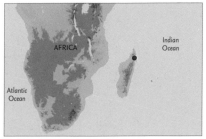

A species is **Critically Endangered** when it faces an extremely high risk of extinction in the wild, based on measurements of population size and/or geographic range and their trends in the past, present and/or future.

© John Dransfield / RBG Kew

The **Forest Coconut**, *Voanioala gerardii*, is listed as 'Critically Endangered' on the IUCN Red List of Threatened Species™. Endemic to Madagascar, this large and highly endangered relative of the coconut is restricted to just a few tiny populations in tropical rainforest in the northeast of the island.

The main threat to the Forest Coconut is exploitation for the edible palm hearts, while the seeds, which are harvested for trade, also have limited dispersal. Most populations are comprised of a few, often scattered, juvenile plants, while of only around ten mature individuals known in the wild, three were felled for palm hearts in 2003. Its extremely small population size puts this palm at high risk of being wiped out by any chance event.

The Forest Coconut is found within an area of the Masoala Peninsula that has recently been declared a national park. No specific conservation measures are currently in place for the species, but genetic and demographic studies have been recommended, while effective protection against exploitation will be vital if the species is to have any chance of survival.

Mulanje Cedarwood

Widdringtonia whytei

NOT EVALUATED	DATA DEFICIENT	LEAST CONCERN	NEAR THREATENED	VULNERABLE	⟨ ENDANGERED ⟩	CRITICALLY ENDANGERED	EXTINCT IN THE WILD	EXTINCT
NE	DD	LC	NT	VU	EN	CR	EW	EX

© Phil Cribb

The **Mulanje Cedarwood**, *Widdringtonia whytei*, is listed as 'Endangered' on the IUCN Red List of Threatened Species™. The national tree of Malawi, this extremely valuable timber tree is endemic to Mount Mulanje, the second highest mountain in southern Africa.

The Mulanje Cedarwood has been heavily exploited in the past, with its wood used as timber and its sawdust distilled to obtain oil for local use as an insecticide. The tree's decline has been somewhat stemmed by a ban on felling, allowing exploitation only of dead trees, but illegal felling and killing of trees continues at an alarming rate.

The Mulanje Mountain Conservation Trust has been set up to provide long-term support for the research and conservation of biological diversity in the Mulanje Mountain Forest Reserve, and the sustainable utilisation of its natural resources. Not only is the Mulanje Cedarwood a national emblem in Malawi, but it is also of critical financial importance to this relatively poor country and, as such, its protection and sustainable use is a national priority.

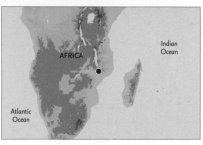

A species is Endangered when it faces a very high risk of extinction in the wild, based on measurements of population size and/or geographic range and their trends in the past, present and/or future.

Wollemi Pine

NOT EVALUATED	DATA DEFICIENT	LEAST CONCERN	NEAR THREATENED	VULNERABLE	ENDANGERED	CRITICALLY ENDANGERED	EXTINCT IN THE WILD	EXTINCT
NE	DD	LC	NT	VU	EN	CR	EW	EX

RED LIST

AUSTRALIA

Tasman Sea

A species is **Critically Endangered** when it faces an **extremely high risk** of extinction in the wild, based on measurements of population size and/or geographic range and their trends in the past, present and/or future.

The **Wollemi Pine**, Wollemia nobilis, is listed as 'Critically Endangered' on the IUCN Red List of Threatened Species™. Considered a 'living fossil', this conifer was only discovered as recently as 1994, and is restricted to just two adjacent populations (not more than 2 km apart) within Wollemi National Park, Australia.

In addition to its extremely limited distribution, the Wollemi Pine has a tiny population, with fewer than 50 mature individuals remaining. The only surviving member of an ancient group dating back to the time of the dinosaurs, this unusual tree may have been undergoing a slow, natural decline for thousands of years. However, its small size and limited range puts the Wollemi Pine at a high risk of extinction from any chance event, such as fire or disease.

The Wollemi Pine is protected in Australia, and the entire population is found within a national park. The population is being monitored, a recovery plan has been drawn up, and propagation trials are underway, in the hope of managing and preserving this unique and intriguing plant.

NOT EVALUATED	DATA DEFICIENT	LEAST CONCERN	NEAR THREATENED	VULNERABLE	ENDANGERED	‹ CRITICALLY ENDANGERED ›	EXTINCT IN THE WILD	EXTINCT
NE	DD	LC	NT	VU	EN	CR	EW	EX

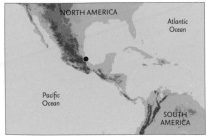

A species is **Critically Endangered** when it faces an extremely high risk of extinction in the wild, based on measurements of population size and/or geographic range and their trends in the past, present and/or future.

© Andrew Vovides

Zamia inermis is listed as 'Critically Endangered' on the IUCN Red List of Threatened Species™. This cycad is endemic to central Veracruz, Mexico, and is known only from one mountain range. There are estimated to be between 300 and 500 individuals remaining in its natural habitat, covering a relatively small area of less than 5 km².

The habitat of Zamia inermis has deteriorated over recent decades due to deforestation for cattle grazing. Forest fires and insecticide use in adjacent sugarcane plantations have also drastically reduced its beetle pollinator population, resulting in poor recruitment of individuals. Illegal collecting has been occurring since before the species was first described in 1983. In its habitat, the species propagates by natural crown breakage, when the crown roots a small distance away from the main stem, which then forms a new crown. Crowns of very old multiheaded plants in cultivation have also been known to break off the main stem.

There is no formal protection for this species. Efforts are being made to conserve the remaining population and to propagate Zamia inermis in an in situ nursery by artificial pollination.

Zamia restrepoi

NOT EVALUATED	DATA DEFICIENT	LEAST CONCERN	NEAR THREATENED	VULNERABLE	ENDANGERED	CRITICALLY ENDANGERED	EXTINCT IN THE WILD	EXTINCT
NE	DD	LC	NT	VU	EN	CR	EW	EX

© ANDERS LINDSTROM

The cycad *Zamia restrepoi* is classified as 'Critically Endangered' on the IUCN Red List of Threatened Species™. Endemic to northern Colombia, the only known location of this species was flooded in 1999 by the completion of the Urra 1 dam that covers over 7,400 hectares of land. The construction of a second dam, Urra 2, further along the river valley has been proposed.

Only a few scattered individuals of this cycad still remain in the wild, and the species is further threatened by local Shamans who use the whole plant as herbal medicine. Collectors, recently interested in obtaining the very last plants of this species, pose an additional risk. Fortunately, the specific pollinators that this cycad relies on to reproduce are still present, and the sporadic production of seeds still occurs.

Zamia restrepoi is currently listed on CITES Appendix I, prohibiting its international trade, as well as being listed as a protected plant within Colombia, but urgent action is required to save this cycad from extinction. The species is slow-growing, and very rare in cultivation both outside and inside its country of origin.

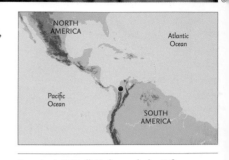

A species is *Critically Endangered* when it faces an extremely high risk of extinction in the wild, based on measurements of population size and/or geographic range and their trends in the past, present and/or future.

NOT EVALUATED	DATA DEFICIENT	LEAST CONCERN	NEAR THREATENED	VULNERABLE	ENDANGERED	CRITICALLY ENDANGERED	EXTINCT IN THE WILD	EXTINCT
NE	DD	LC	NT	VU	EN	CR	EW	EX

© Dr Stephanos Diamandis

The fungus, Zeus olympius, is found only on twigs of a few trees of Bosnian pine from Mount Olympus in northern Greece, where it is apparently endemic: despite searches, it has not been found elsewhere or on other plants. Although not yet officially listed on the IUCN Red List of Threatened Species™, its conservation status is potentially 'Critically Endangered'.

Zeus olympius produces circular fruitbodies below the bark and a blackened protective crust. When ripe, the fruitbody expands, breaking through the bark, and splitting the crust into small black teeth, thereby revealing the fertile layer – an orange- or golden-coloured disc about 5 mm in diameter. To reporoduce, its spores are ejected from this surface and dispersed by wind. Although possibly parasitic, it seems to pose no serious threat to the pine.

The area of the national park with known colonies of this fungus is very small, and the species may be particularly vulnerable as it is found near picnic sites and adjacent to recreational trails. Severe forest fires, natural or otherwise, are a real threat.

A species is **Critically Endangered** when it faces an extremely high risk of extinction in the wild, based on measurements of population size and/or geographic range and their trends in the past, present and/or future.

Tequila Splitfin

Zoogoneticus tequila

NOT EVALUATED	DATA DEFICIENT	LEAST CONCERN	NEAR THREATENED	VULNERABLE	ENDANGERED	CRITICALLY ENDANGERED	EXTINCT IN THE WILD	EXTINCT
NE	DD	LC	NT	VU	EN	CR	EW	EX

© ROMAN SLABOCH

The **Tequila Splitfin**, *Zoogoneticus tequila*, is listed as 'Critically Endangered' on the IUCN Red List of Threatened Species™. Believed 'Extinct in the Wild' until its rediscovery in 2003, this fish, named after the Tequila Volcano to the north of its range, is known from just one tiny pool at Teuchitlán, Mexico.

The remaining wild population of the Tequila Splitfin consists of fewer than 500 individuals, less than 50 of which are mature adults. Its original decline in the Teuchitlán River has been attributed to the introduction of exotic fish and habitat degradation due to the building of a dam.

Native fish populations have also been fragmented, as the springs they were confined to have been turned into spas. In addition, the Tequila Splitfin is under threat from intensive collection for the aquarium trade, and by pollution and water extraction in its one remaining pool.

The Tequila Splitfin is kept in captivity in North America and Europe, and a breeding programme has been recommended so that the species can be reintroduced into further, larger pools. Legal protection is also needed, as is a control in international trade.

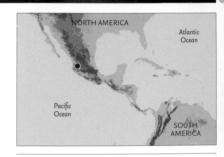

A species is **Critically Endangered** when it faces an extremely high risk of extinction in the wild, based on measurements of population size and/or geographic range and their trends in the past, present and/or future.

NOT EVALUATED	DATA DEFICIENT	LEAST CONCERN	NEAR THREATENED	⟨ VULNERABLE ⟩	ENDANGERED	CRITICALLY ENDANGERED	EXTINCT IN THE WILD	EXTINCT
NE	DD	LC	NT	VU	EN	CR	EW	EX

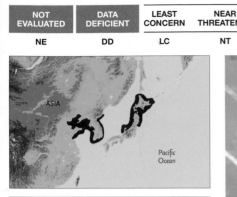

A species is **Vulnerable** when it faces a high risk of extinction in the wild, based on measurements of population size and/or geographic range and their trends in the past, present and/or future.

© F.T. Short, SeagrassNet

Zostera caespitosa is listed as 'Vulnerable' on the IUCN Red List of Threatened Species™. This seagrass is known from a few locations on the northern coast of China, the Korean Peninsula and in northern Japan. *Zostera caespitosa* has a shallow depth range (2–11 metres) as well as a limited and patchy distribution.

Zostera caespitosa has experienced recent declines in many parts of its range. It is sensitive to pollution and to reductions in water quality, which limit light availability. Habitat loss is also a major threat to this species, since within its range in Japan and China there is industrial development in coastal regions, land reclamation, and disturbance by fish trawling and other fishing practices. In China, heavy use of the coastal zone for kelp aquaculture affects this seagrass. In South Korea, coastal eutrophication and land reclamation impact the health of *Zostera caespitosa*.

Very little of the area where this threatened seagrass is found receives any kind of protection. In South Korea, *Zostera caespitosa* is protected by The Marine Ecosystem Conservation and Management Act.

Zyzomys pedunculatus

NOT EVALUATED	DATA DEFICIENT	LEAST CONCERN	NEAR THREATENED	VULNERABLE	ENDANGERED	CRITICALLY ENDANGERED	EXTINCT IN THE WILD	EXTINCT
NE	DD	LC	NT	VU	EN	CR	EW	EX

The **Central Rock Rat**, *Zyzomys pedunculatus*, is listed as 'Critically Endangered' on the IUCN Red List of Threatened Species™. This enigmatic rat is known from 14 locations in the West MacDonnell Range National Park in Australia's Northern Territory, but has not been seen since drought and wildfire struck in 2002, and may be 'Extinct in the Wild'.

Prior to its rediscovery in 1996, the Central Rock Rat had not been seen since 1960 and was believed to be extinct. Thus, its recent disappearance is not entirely unprecedented, and it is quite possible that the species remains undetected at low numbers. Current potential threats include fire, the spread of non-native grasses, Dingo predation, and habitat degradation due to grazing by feral herbivores.

Field surveys are urgently needed to try and relocate this species in the wild. Fortunately, captive breeding is underway at Perth Zoo and Alice Springs Desert Park, and the feasibility of reintroducing populations to parts of its range is being considered.

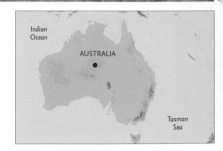

A species is Critically Endangered when it faces an extremely high risk of extinction in the wild, based on measurements of population size and/or geographic range and their trends in the past, present and/or future.

Australian EEZ Australian Exclusive Economic Zone (also known as Economic Exclusion Zone) – A designated area extending for 200 nautical miles (370 km) from the coastline around the continent, within which the state has rights and responsibilities over the exploration, exploitation, conservation and management of all marine resources.

Bern Convention – The Convention on the Conservation of European Wildlife and Natural Habitats (sometimes shortened to the 'Bern Convention') is an international agreement between many European states, and also some states in northern Africa. It came into force in 1982, with the aim of conserving wild flora and fauna, and promoting cooperation between signatory states.

BGCI Botanic Gardens Conservation International – An international organization which aims to ensure the conservation of threatened plants around the world by means of securing specimens within botanic garden collections.

CITES Convention on International Trade in Endangered Species of Wild Flora and Fauna – A voluntary international agreement between governments, whose aim is to ensure that any trade in wild flora and fauna, and products derived from them, does not threaten the survival of the species. Appendix I lists species that are threatened with extinction, and Appendix II lists those whose trade must be controlled in order to safeguard their future.

Convention on the Conservation of Migratory Species of Wild Animals (also known as CMS or Bonn Convention – An intergovernmental treaty established by the United Nations Environment Programme (UNEP), with the aim of conserving migratory species throughout their range. Since it came into force in 1983, legally binding agreements have been drawn up to ensure the protection of various terrestrial, marine and avian species within the member states.

Crop Wild Relative project – An initiative, led by the Global Crop Diversity Trust, involving the collection of species of wild plants that are genetically related to domesticated plants. Driven by the need to produce crops that are adapted to climate change, and therefore to ensure food security, the project will also ensure the conservation of the collected species themselves.

Fauna & Flora International (FFI) – An international conservation organization which, since its inception in 1903, has established various projects around the world aimed at protecting plant and animal species and their habitats.

Global Trees Campaign – A joint initiative between Fauna & Flora International and Botanic Gardens Conservation International, aimed at saving threatened tree species around the world.

International Agreement on the Conservation of Polar Bears – A treaty negotiated in 1973 by the five nations with the largest polar bear populations (Canada, Greenland, Norway, the United States and the former Soviet Union) aimed at restricting the hunting of polar bears and protecting their habitats.

International Council for the Exploration of the Sea (ICES) – An international scientific community concerned with coordinating and promoting marine research in the North Atlantic Ocean. Inaugurated in 1902, its member states include all those bordering the North Atlantic Ocean and the Baltic Sea.

International Whaling Commission (IWC) – Set up to implement the 1946 International Convention for the Regulation of Whaling, the objective of this intergovernmental body is to regulate whaling activities, protect threatened species and promote scientific research throughout the world.

IUCN International Union for Conservation of Nature – Founded in 1948, this environmental network of government and non-governmental organizations is dedicated to providing conservation and sustainability at both global and local levels.

IUCN Red List of Threatened Species™ – A list, developed and maintained by IUCN, that evaluates the conservation status of threatened plant and animal species around the world. The goals of the list are to identify and document those species in most need of conservation, and to provide a global index of the state of change of biodiversity.

Marine Protected Area (MPA) –
A designated area of ocean, sea or
estuary within which the environment
and biodiversity, and sometimes
cultural resources, are protected
by legislation from damage or
disturbance by human activity.

Millennium Seed Bank project – A project
to collect and store seeds from plant
species around the world. Coordinated
by the Royal Botanic Gardens at Kew,
UK, the emphasis is on those species
most at risk from both climate change
and the effects of human activity. The
seeds are banked at Kew's own seed
bank and also in various partner seed
banks in other parts of the world.

Natura 2000 network – A network
of protected areas created to help
conserve the most seriously threatened
habitats and species across Europe.
The areas have been designated in
response to the Habitats Directive
and the Birds Directive, both pieces of
legislation adopted by the European
Union in 1992 and 1979 respectively.

**NGO Non-governmental
organization** – A body which operates
independently of any government.

**North Atlantic Salmon Conservation
Organization (NASCO)** – An
organization established in 1984 by
an intergovernmental Convention
with the objective of managing and
conserving Atlantic salmon stocks.

Persistent organic pollutants (POPs)
– Organic chemical compounds that,
once released into the environment,
remain there for long periods of
time, causing a hazard to wildlife.

Projeto Tamanduá – A Brazilian
project concerned with the
conservation of anteater species.

TRAFFIC – A wildlife trade monitoring
network established in 1976 to
promote sustainable practices within
the trade of wild plants and animals.
There are nine regional programmes
including TRAFFIC East/Southern
Africa, which was founded in 1991.

Working for Water Programme –
A government-led initiative aimed
at controlling the spread of invasive
alien plant species in South Africa.

World Heritage Site – A cultural, historical
or natural site designated by the United
Nations Educational, Scientific and
Cultural Organization (UNESCO) as
being of global significance and a priority
for protection and conservation.

This book is inspired by the IUCN Species of the Day 2010 web initiative, and therefore owes special thanks to all the contributors, particularly the project team who, managed by Rachel Roberts, Species Survival Commission (SSC), and assisted by Kathryn Pintus, IUCN Global Species Programme, compiled all the profiles of this book.

Much credit is due to all the IUCN SSC Specialist Groups who nominated species, reviewed the text and helped source images, and to all the photographers who kindly supplied the vast diversity of images.

We would especially like to thank:

IUCN Global Species Programme staff,
including:
Biodiversity Assessment Unit
Communications Unit (Lynne Labanne)
Finance and Administration
Freshwater Biodiversity Unit
Management (Jane Smart)
Marine Biodiversity Unit
Network Support Officers
Red List Unit
Species Information Service Unit

Species Survival Commission Chair's Office
Simon Stuart and Katharine Holmes

Additional Contributors:
Australian Marine Conservation Society
BirdLife International (Nigel Collar)
Durrell Wildlife Conservation Trust
   (Richard Young)
Isha Marquez
IUCN Publications (Deborah Murith)
Solertium
Studiohope Limited
UNEP
Zoological Society of London
   (EDGE Team, Ian Stephen)

ARKive
ARKive Media Research Team
ARKive Research Managers
   (Michelle Lindley, Lucie Muir, Verity Pitts,)
ARKive Species Text Authors
   (Andrew Letten, Alex Royan, Elisabeth Shaw)

IUCN SSC Specialist Groups
African Elephant
African Rhino
Afrotheria
Amphibian
Anteater, Armadillo and Sloth
Antelope
Asian Rhino
Asian Wild Cattle
Australasian Marsupial and Monotreme
Bat
Bear
Bison
Bryophyte
Bumblebee
Cactus and Succulent
Canid
Caprinae
Cat
Cetacean
Chameleon
China Plant
Conifer
Coral
Crocodile
Crop Wild Relative
Cup-fungi, Truffles and Allies
Cycad
Deer
Dragonfly
Eastern Africa Plant Red List Authority
Equid
Freshwater Crab and Crayfish
Freshwater Fish

Galliformes
Global Tree
Grasshopper
Grouper and Wrasse
Hippo
Hyeana
Iguana
Lagomorph
Lichen
Macaronesian Island Plant
Madagascar Plant
Marine Turtle
Mascarene Plant
Mediterranean Island Plant
Medicinal Plant
Mollusc
Non-volant Small Mammal Red List Authority
New World Marsupial
Otter
Palm
Peccary
Pelican
Pinniped
Primate
Salmon
Seagrass Red List Authority
Sea Snake
Shark
Sirenia
Small Carnivore
Southern African Plant
South American Camelid
South Asian Invertebrate
Sturgeon
Tapir
Temperate South American Plant
Threatened Waterfowl
Tortoise and Freshwater Turtle
Wild Pig
Wolf

IUCN, International Union for Conservation of Nature, helps the world find pragmatic solutions to our most pressing environment and development challenges by supporting scientific research; managing field projects all over the world; and bringing governments, NGOs, the UN, international conventions and companies together to develop policy, laws and best practice.

The world's oldest and largest global environmental network, IUCN is a democratic membership union with more than 1,000 government and NGO member organizations, and almost 11,000 volunteer scientists and experts in some 160 countries. IUCN's work is supported by over 1,000 professional staff in 60 offices and hundreds of partners in public, NGO and private sectors around the world. IUCN's headquarters are located in Gland, near Geneva, in Switzerland. www.iucn.org/species

The IUCN Species Survival Commission (SSC) is the largest of IUCN's six volunteer commissions with a global membership of over 7,000 experts. SSC advises IUCN and its members on the wide range of technical and scientific aspects of species conservation and is dedicated to securing a future for biodiversity. SSC has a significant input into the international agreements dealing with biodiversity conservation. www.iucn.org/species

The IUCN Red List of Threatened Species™ (or The IUCN Red List) is the world's most comprehensive information source on the global conservation status of plant and animal species. It is based on an objective system for assessing the risk of extinction of a species should no conservation action be taken.

Species are assigned to one of eight categories of threat based on whether they meet criteria linked to population trend, population size and structure, and geographic range. Species listed as Critically Endangered, Endangered or Vulnerable are collectively described as 'Threatened'.

The IUCN Red List is not just a register of names and associated threat categories. It is a rich compendium of information on the threats to the species, their ecological requirements, where they live, and information on conservation actions that can be used to reduce or prevent extinctions.

ARKive is an initiative of Wildscreen, an international charity working to promote the public understanding and appreciation of the world's biodiversity and the need for its conservation through the power of wildlife imagery. ARKive is gathering together the very best films and photographs of the world's species into one centralised digital library, to create a stunning audio-visual record of life on Earth. ARKive's immediate priority is to compile audio-visual profiles for the c. 18,000 animals, plants and fungi featured on The IUCN Red List of Threatened Species. www.arkive.org